O9-BUC-309

FUZZY THINKING

ALSO BY BART KOSKO

Neural Networks and Fuzzy Systems (textbook)
Neural Networks for Signal Processing (textbook)

FUZZY THINKING

THE NEW SCIENCE OF FUZZY LOGIC

BART KOSKO

NEW YORK

Figures 11.2, 11.3, and 11.4 drawn after an illustration from B. Kosko, "Adaptive Bidirectional Associative Memories," *Applied Optics* 26, no. 23 (1987), pp. 4947–60. Figure 11.10 drawn after an illustration from B. Kosko, "Fuzzy Cognitive Maps," *International Journal of Man-Machine Studies* 24 (January 1986), pp. 65–75. Figure 11.14 drawn after an illustration from B. Kosko, *Neural Networks and Fuzzy Systems: A Dynamical Systems Approach to Machine Intelligence,* Englewood Cliffs, NJ: Prentice Hall, 1992. Figure 11.15 drawn after an illustration from R. Taber, "Knowledge Processing with Fuzzy Cognitive Maps," *Expert Systems with Applications* 2, no. 1 (1991), pp. 83–87.

Copyright © 1993 Bart Kosko, Ph.D.

All rights reserved. No part of this book may be used or reproduced in any manner whatsoever without written permission of the Publisher. Printed in the United States of America. For information address Hyperion, 114 Fifth Avenue, New York, New York 10011.

Library of Congress Cataloging-in-Publication Data
Kosko, Bart.
Fuzzy thinking : the new science of fuzzy logic/Bart Kosko.—
1st ed.
p. cm.
ISBN 0-7868-8021-X
1. Logic. 2. Philosophy and science. 3. Fuzzy systems.
I. Title. II. Title: Fuzzy logic.
BC108.K59 1993 160-dc20 92-42019 CIP

FIRST PAPERBACK EDITION

10 9 8 7 6 5 4 3 2 1

For the young men and women
who stick with their training
while their youth calls.
It's hard, it will get harder,
but it turns the world.

Acknowledgments

This book grew out of years of argument. I thank my many friends and students and colleagues for their time and ideas and for the chance to debate them.

Special thanks go to Leslie Meredith and Sheldon Teitelbaum, who first got me to put the arguments on paper. Most of all I want to thank Tom Miller, senior editor at Hyperion, who had the vision and courage to support the evolving manuscript.

CONTENTS

PART THREE

THE FUZZY PRESENT

PREFACE

One day I learned that science was not true. I do not recall the day but I recall the moment. The God of the twentieth century was no longer God.

There was a mistake and everyone in science seemed to make it. They said that all things were true or false. They were not always sure which things were true and which were false. But they were sure that all the things were either true or false. They could say whether grass is green or whether atoms vibrate or whether the number of lakes in Maine is an even or odd number. The truth of these claims had the same truth as claims about math or logic. They were true all or none, white or black, 1 or 0.

In fact, they were matters of degree. All facts were matters of degree. The facts were always *fuzzy* or vague or inexact to some degree. Only math was black and white and it was just an artificial system of rules and symbols. Science treated the gray or fuzzy facts as if they were the black-white facts of math. Yet no one had put forth a single fact about the world that was 100% true or 100% false. They just said they all were.

That was the mistake and with it came a new level of doubt. Scientists could err at the level of logic and math. And they could maintain that error with all the pomp and intolerance of a religious cult.

I pursued gray truth in my graduate training in math and electrical engineering and machine intelligence. At first I worked with symbols on abstract math theorems. This made the gray or fuzzy world view seem like a dry exercise in a math textbook. Then I started to teach the subject and to combine the math with real applications of the fuzzy ideas. Students were quick to learn how to paint gray pictures of a gray world. Some students built real fuzzy systems or software packages and some of them patented their ideas and some of them went off to sell their wares or to start their own companies. In Japan engineers designed the first fuzzy "smart" commercial products. Soon there would be fuzzy camcorders and washing machines and microwave ovens and carburetors and hundreds of other smart products.

The applications showed that the fuzzy world view extended beyond the journal paper and textbook and classroom. The rapid spread of fuzzy ideas in the Far East and the opposition to them in the West showed even more. The fuzzy world view was a world view. It extended as much to culture and philosophy as it did to science and math. It reached back to thinkers as diverse as Aristotle and the Buddha. It reached forward to how we argue over law and abortion and the nature of intelligence and on to how someday we might make our peace with machines that can beat us in any race.

This book is my statement of the fuzzy world view. At the core is the paradigm shift from the black and white to the gray—from bivalence to multivalence. I tell the tale with a mix of science and philosophy and history and with episodes from my own fuzzy experience. The point was not to write a text on fuzzy logic. I already did that and it takes too many equations. The point was to show the fuzzy world view at work in the mind and in the flesh. To do that you have to have lived the field and fought the fights. You have to have doubted the God of science and felt a little of Her wrath.

Into every tidy scheme for arranging the pattern of human life, it is necessary to inject a certain dose of anarchism.

BERTRAND RUSSELL
SCEPTICAL ESSAYS

I

The Fuzzy Principle

Everything Is a Matter of Degree

1

Shades of Gray

So far as the laws of mathematics refer to reality, they are not certain. And so far as they are certain, they do not refer to reality.

Albert Einstein
Geometry and Experience

Fuzzy theory is wrong, wrong, and pernicious. What we need is more logical thinking, not less. The danger of fuzzy logic is that it will encourage the sort of imprecise thinking that has brought us so much trouble. Fuzzy logic is the cocaine of science.

Professor William Kahan
University of California at Berkeley

"Fuzzification" is a kind of scientific permissiveness. It tends to result in socially appealing slogans unaccompanied by the discipline of hard scientific work and patient observation.

Professor Rudolf Kalman
University of Florida at Gainesville

Fuzziness is probability in disguise. I can design a controller with probability that could do the same thing that you could do with fuzzy logic.

Professor Myron Tribus, on hearing of the fuzzy-logic control of the Sendai subway system
IEEE Institute, May 1988

Hold an apple in your hand. Is it an apple? Yes. The object in your hand belongs to the clumps of space-time we call the set of apples—all apples anywhere ever. Now take a bite, chew it, swallow it. Let your digestive tract take apart the apple's molecules. Is the object in your hand still an apple? Yes or no? Take another bite. Is the new object still an apple? Take another bite, and so on down to void.

The apple changes from thing to nonthing, to nothing. But where does it cross the line from apple to nonapple? When you hold half an apple in your hand, the apple is as much there as not. The half apple foils all-or-none descriptions. The half apple is a *fuzzy* apple, the gray between the black and the white. *Fuzziness is grayness.*

Seventeenth-century philosopher René Descartes pondered a plug of beeswax one evening in front of his fireplace. He thumped the wax and heard its dull sound, smelled its honey fragrance, felt its smooth surface and cool temperature, tried to look through its milky texture. Then he held the wax plug near the fire. The hard white plug softened, warmed, stretched, lost its odor, became clear, became liquid, flowed. Some of it dripped onto the hot brick hearth and sizzled and boiled away into the atmosphere.

Where did the wax go? When did it change from wax plug to non–wax plug? Where did its identity lie? In the plug? On the hearth? In between?

We face the same questions every day when we look in the mirror. The face and hair and teeth and skin have changed. They even change slightly, at the molecular level, as we look. We pass slowly from self to a new nonself as we grow old one cell at a time, a molecule at a time. Molecules assemble and disassemble an atom or two at a time. Atoms assemble and disassemble a quark at a time. Quarks, our current smallest units of matter, assemble and disassemble in ways we do not yet know. The division of matter may go all the way down to Leibniz's monads, infinitesimal but intelligent points of existence.

All around us things change their identities. The atoms that make up the universe swirl and collide and keep swirling and colliding. Everything is in flux. Everything flows. The universe unfolds as a river runs. The cosmological fluid seems to obey Einstein's laws of general relativity in the large and seems to obey the laws of quantum mechanics in the small and obeys we do not know what in between.

Things flow smoothly to nonthings. The atoms in our fingertips swirl into the atoms in the air. There are finger atoms and nonfinger atoms. And there are the atoms in between, the atoms to some degree both finger atoms and air atoms and to some degree neither. A rose is a rose is a nonrose when its molecules change. The finger shades into the hand, the hand shades into the wrist, the wrist into the arm. Earth's atmosphere shades into space. The mountain crumbles into a hill and in time crumbles into a plain. The growing human embryo passes into a living human being and a living brain decays into death.

We can put black-and-white labels on these things. But the labels will pass from accurate to inaccurate as the things change. Language ties a string between a word and the thing it stands for. When the thing changes to a nonthing, the string stretches or breaks or tangles with other strings. "House" stands for a house even after the house falls apart or burns. Our world of words soon looks like a fishing boat that drifts with thousands of tangled and broken lines.

We know things change. Science reveals a world of jagged edges and quantities that vary smoothly. More precision does not take the gray out of things—it pins down the gray. Medical advances have not made it easier to draw the line between life and not-life at birth or at death. If we described the Earth's atmosphere molecule by molecule, we would still find no line that divides the atmosphere from space. Detailed maps of the surface of Earth and Mars and the Moon do not tell us where hills end and mountains begin.

Yet in much of our science, math, logic, and culture we have assumed a world of blacks and whites that does not change. Every statement is true or false. Every molecule in the cosmos belongs to your finger or not. Every law, statute, and club rule applies to you or not. The digital computer, with its high-speed binary strings of 1s and 0s, stands as the emblem of the black and white and its triumph over the scientific mind.

This faith in the black and the white, this *bivalence*, reaches back in the West to at least the ancient Greeks. Democritus reduced the universe to atoms and void. Plato filled his world with the pure forms of redness and rightness and triangularity. Aristotle took time off from training his pupil Alexander the Great to write down what he felt were the black-and-white laws of logic, laws that scientists and mathematicians still use to describe and discuss the gray universe.

Aristotle's binary logic came down to one law: A OR not-A. Either this or not this. The sky is blue or not blue. It can't be both blue and not blue. It can't be A AND not-A. Aristotle's "law" defined what was philosophically correct for over two thousand years.

The binary faith has always faced doubt. It has always led to its own critical response, a sort of logical and philosophical underground. The Buddha lived in India five centuries before Jesus and almost two centuries before Aristotle. The first step in his belief system was to break through the black-and-white world of words, pierce the bivalent veil and see the world as it is, see it filled with "contradictions," with things and not-things, with roses that are both red and not red, with A AND not-A.

You find this fuzzy or gray theme in Eastern belief systems old and new, from Lao-tze's Taoism to the modern Zen in Japan. Either-or versus contradiction. A OR not-A versus A AND not-A. Aristotle versus the Buddha.

The Greeks called their dissenters "sophists." Today we call sophistry reasoning we find faulty or silly. When one day in his Academy Plato defined man as a featherless biped, the next day a sophist student walked into class and held out to Plato a plucked chicken. Zeno picked a grain of sand from a sand heap and asked whether the heap was still a heap. Zeno could never find that sand grain that changed the heap to a nonheap, that took it from A to not-A. As he picked out more sand grains he seemed to get a heap and a nonheap, A AND not-A. The liar from Crete said that all Cretans are liars and asked if he lied. If he lied, then he did not. And if he did not lie, then he did. He seemed to lie and not lie at the same time. Modern philosophers, like Descartes, have brooded over the nature of identity and have searched in vain for the common substance that passes from wax plug to non–wax plug. David Hume saw the self dissolve into a nonself bundle of sensations. Werner Heisenberg showed physicists that not all scientific statements are true or false. Many, if not most, statements are indeterminate, uncertain, gray—fuzzy. Logician Bertrand Russell found the Cretan's liar paradox at the foundations of modern math. Mathematicians and philosophers have since tried to patch and gerrymander those black-and-white foundations to get rid of the gray paradoxes. But the paradoxes and the brooding about them remain.

I brooded about grayness too. It led me from philosophy to mathematics to electrical engineering. I picked up degrees along the way and in time ended up teaching electrical engineering at the University of Southern California, where I had started out as an undergraduate in music composition. I ran across Einstein's quote about math mismatching reality when I was 21 and sat in a USC philosophy classroom. It shocked me to see Einstein doubt the very math framework of black-and-white science that he had helped build: "So far as the laws of mathematics refer to reality, they are not certain. And so far as they are certain, they do not refer to reality." So Einstein brooded about grayness too.

I skimmed Sir A. J. Ayer's anthology on *logical positivism*, the dominant philosophy of science in this century. Positivism demands evidence, factual or math evidence, as a security guard demands *positive* ID, not just your say-so. Logical positivism holds that if you cannot test or mathematically prove what you say, you have said nothing. Positivism works out well for scientists and mathematicians, since it allows only them to speak. Everyone else utters "meaningless" statements about the world and life and morals and beauty. Problems of God and metaphysics and goodness and value reduce to mere "pseudo-problems," questions asked by those whom language has misled, those who do not know what counts as answers. Logical positivist Moritz Schlick ended one of his essays in the anthology on just this point:

> Philosophical writers will long continue to discuss the old pseudo-questions. But in the end they will no longer be listened to. They will come to resemble actors who continue to play for some time before noticing that the audience has slowly departed. Then it will no longer be necessary to speak of "philosophical problems."

Every philosopher you ask will attack logical positivism, either on details or on some general principle, but it remains the working philosophy of modern science, medicine, and engineering. Logical positivism hands the future to scientists. It also hands them much of the present.

Einstein's quote punched through the black-and-white world of science and math. It sounded the positivists' lament. I read the quote over and over and found myself slowly becoming a fuzzy logical positivist. The world of math does not fit the world it describes. The two worlds differ, one artificial and the other real, one neat and the other messy. It takes faith in language, a dose of make-believe, to make the two worlds match.

I called this the *mismatch problem*: *The world is gray but science is black and white*. We talk in zeroes and ones but the truth lies in between. Fuzzy world, nonfuzzy description. The statements of formal logic and computer programming are all true or all false, 1 or 0. But statements about the world differ.

Statements of fact are not all true or all false. Their truth lies between total truth and total falsehood, between 1 and 0. They are not bivalent but *multi*valent, gray, fuzzy. These statements are not just tentative, they are imprecise and vague. The logical statement "Two equals two" and the math statement "2 + 2 = 4" are precise and 100% true—true, as philosophers say, "in all possible universes," even though philosophers have seen only one. But that does not affect how atoms swirl or how universes expand or how a strawberry tastes or how a face slap feels. We can never prove 100% true a scientific statement or claim of fact like "The moon shines" or "Grass is green" or "$e = mc^2$." Fresh evidence may topple any scientific belief, and objects of belief differ only approximately from their opposites. The blade of green grass turns brown. In the next instant the moon may stop shining and burst into flames or fall into the Earth or break a cherished law of science and collapse into a black hole or a ball of cheese.

Laws of science are not laws at all. They are not laws in the sense of logical laws like two plus two equals four. Logic does not legislate them. Laws of science state tendencies we have recently observed in our corner of the universe. The best you can say about them is so far, so good. In the next instant every "law" of science may change. Their truth is a matter of degree and is always up for grabs. Yet the language of science, the language of math and logic and computer programming, is black and white. It deals only with statements that are 100% true or 100% false. Math talk differs in kind from science talk. But scientists talk it anyway.

I thought scientists and philosophers would see the mismatch problem as the central philosophical problem of modern science. But they seemed to ignore it. Einstein called it out as a quotable quote, an ironic throw-away line in the midst of the business of science. A rare philosopher might cite Einstein's quote to bring out the positivist's favorite split, between words and objects, logic and fact. But I found no one wrestling with the mismatch problem and for a reason: they assumed the match too.

Philosophers assumed the world was black and white, bivalent, just like the words and math they used to describe it. After all these years and all that training they still took orders from Aristotle and did not question them. In theory they could tell matters of logic from matters of fact. In practice they ignored this split and treated the messy matters of fact as if they were neat matters of logic. They did this for two reasons: First, it was easy. Second, it was habit. They used the same artificial language to talk about matters of logic and matters of fact. They described math and the world with the same black-white "symbolic logic" that Aristotle set in motion over two thousand years ago.

And I used it too. In gym class the toughest guys do the most pushups or run the fastest mile or punch hardest. In a modern philosophy class they find the shortest proofs to the theorems of symbolic logic. The same holds in science. The more math an author throws at a problem, the less her audience understands her and the more they respect her. Your skill at logic and math places you in the pecking order of science. I competed with classmates and professors to move up in that pecking order. I saw formal logic as the flame the philosophers kept. But Einstein's quote kept eating through my faith in logic and science.

At first my doubt had no focus. I had no alternative to the black-white formalisms of science and logic. What else could there be? To stray from logic and science struck me as the nonlogic and nonsense of Eastern mysticism, the full-lotus stuff a young scientist makes jokes about when he eats in a Chinese restaurant.

Even Einstein had no alternatives to bivalence. Instead he and the league of scientists added a new theory to the old theory of bivalence. They added the theory of probability, the mathematical theory of "chance" or "randomness." The idea is that every event has a number attached to it, the probability that the event will occur. The event might be the flip of a coin. There is a probability that the coin comes up heads and a probability that it comes up tails. The coin comes up heads or tails and the two probabilities add up to one. In general the probability that an event occurs and the probability that it does not occur add up to one. That's probability theory. Event numbers add up to one and the events are black and white, they occur all or none.

Probability did not alter or even challenge the black-white picture of the world. It just showed how to gamble on it and in it. Aristotle's law of A OR not-A always holds in probability. The new physicists saw probability wherever they looked. But Einstein did not feel comfortable with it. That is what he meant when he said that "God does not play dice." Quantum mechanics, the physics of subatomic events, suggested otherwise. The universe seemed nothing but probability.

I found comfort in the math of probability but not in the very idea of it. What is probability? What kind of thing is it? What does it look like? How do you measure it? How do you test a probability claim? I hold a coin in my hand and claim that it is "fair"—the coin has a 50% "chance" or "probability" of coming up heads. Then I flip the coin and it comes up heads. Does that confirm the claim? If it does, it also confirms the claim that the heads probability is 55% or 90% or even 100%. We could even count the head flip as evidence for the claim that the heads probability is 45% or 10% or even 0.0000000001%. Just the "luck" of the draw. A probability experiment can go either way and you do not know which way. If I hold the white chess pawn in my hand behind my back and ask you to guess which hand holds it, I know which way the experiment goes but you must guess, estimate, calculate odds. For you the probability that my right hand holds the pawn is "real" or makes sense. For me it is an illusion. I know the outcome with "certainty."

Probability evaporates with increased information. Information up, probability down. The laws of physics determine whether the coin comes up heads or tails. To a supersmart, supersensitive, supercomputing being all probability experiments are illusions. So maybe there is no probability. Maybe there is something else, maybe something fuzzy, that we sometimes call probability but that exists in the nature of things and nonthings, or in the relations between them.

I looked for the answer as to how probability differed from fuzziness but could not find it, because at that time I did not know what fuzziness was. I had not seen the math of fuzziness. In time I wrote my Ph.D. dissertation on fuzzy math to help me see and that still was not enough. I wanted to draw a line in the math sand between fuzziness and probability. But in my heart I suspected one contained the other. Critics said that all the time: fuzziness is probability in disguise.

I suspected it was the other way around. Since the days of ancient Sumeria, men and women have used probability words to refer to complex patterns of behavior in the environment and in society—whether it will rain, whether the beer will spoil, whether the hunters will find a deer, whether the other village will attack, whether the wife will get pregnant. Modern scientists have carried on this tradition and they did not question it. Instead they put the probability talk in math and exalted it in all the sciences. That made me even more suspicious. The probability that the bowman's arrow hits the deer does not lie in the arrow or the deer. If it lies anywhere it lies in his mind, his brain state, or in ours. Math or no math, I wondered, how wise is it to build these brain states into the foundations of quantum mechanics, the theory of subatomic particles, into the most fundamental descriptions of the universe?

These were problems with the idea of probability. But there was still the first problem of applying probability to the real world: Probability did not solve the mismatch problem. It compounded the problem. It piled a new theory on top of the black-white theory of bivalence. And that came with a price: the world filled with the new phantom of "randomness," the concept mathematicians for years have tried to define.

Probability deals with blacks and whites—heads or tails, success or failure, in the box or out. It puts odds on these precise events. It does not remove their precision, their bivalent flavor. Indeed scientists round off gray things to black or white things before they apply probability. The electron orbits the atomic nucleus or not. The cell cluster turns cancerous or not. The lion catches the gazelle or not. The customer waits in line or not. The cloud is nimbus or not. The star belongs to the galaxy or not. The universe is opened or closed. At the conceptual level probability theory has filled the universe with the undefined, unobservable gas of "randomness." In practice it has made scientists draw more, not fewer, lines between things and nonthings. Probability has turned modern science into a truth casino.

The practice began almost a half millennium ago when scientists worked out the first probability math from examples of gambling, from games of chance with artificially precise rules and cutoff lines. Two or three centuries later scientists applied probability to disease and death statistics of city populations to give mathematical rebirth to the insurance industry. Either you had the disease or not, were married or not, were over 20 or not, were over the poverty line or not, were dead or not. Today commanders launch attacks or not depending on kill probabilities. Probability has proved a powerful tool for social prediction and control. But I could not see how it softened the mismatch between logic and fact.

Say you park your car in a parking lot with 100 painted parking spaces. The probability approach assumes you park in one parking space and each space has some probability that you will park in it. All these parking-space probabilities add up to 100%. If the parking lot is full, there is zero probability that you will park in it. If there is only one empty parking space, say the thirty-fourth space, you will park there with 100% probability. If the parking lot is empty, and if we know nothing else about the parking lot, you have the same slim chance, 1%, of parking in any one of the parking spaces.

The probability approach assumes parking in a space is a neat and bivalent affair. You park in the space or not, all or none, in or out. A walk through a real parking lot shows otherwise. Cars crowd into narrow spaces and at angles. One car hogs a space and a half and sets a precedent for the cars that follow. To apply the probability model we have to round off and say one car per space.

Up close things are fuzzy. Borders are inexact and things coexist with nonthings. You may park your car 90% in the thirty-fourth space and 10% in the space to the right of it, the thirty-fifth space. Then the statement "I parked in the thirty-fourth parking space" is not all true and the statement "I did not park in the thirty-fourth parking space" is not all false. To a large degree you parked in the thirty-fourth space and to a lesser degree you did not. To some degree you parked in *all* the spaces. But most of those were zero degrees. This claim is fuzzy and yet more accurate. It better approximates the "fact" that you parked in the thirty-fourth parking space.

I found a different fuzzy example as I sat in the philosophy classroom. The professor asked a question. I do not remember the question but I remember resenting the suggestion that either you know the answer or not and, if you know it, you raise your hand and in due course state the correct answer. Children first meet this bivalent filter when they go to preschool or kindergarten. Know it or not. Hands up or down. Answer right or wrong. Put up or shut up. I felt I knew a partial answer to the question. I did not know the answer all or none with a probability to decide which. I had just been studying multivalued or fuzzy logic and so I thought it made sense to raise my hand only part way up to show the degree to which I knew the answer. The innovation failed and the professor called on me to answer all or none.

I have since used this trick on audiences to show them a "real" fuzzy set. How many of you are male? Raise your hands. Male hands go up and female hands stay down. That gives a set and it is not fuzzy. Aristotle's A OR not-A still holds. How many of you are female? Raise your hands. The reverse happens and again the audience splits into two black-white sets, males and nonmales or females and nonfemales.

Then comes a harder question: How many of you are satisfied with your jobs? The hands bob up and down and soon come to rest with most elbows bent. A confident few point their arms straight up or do not raise them at all. Most persons are in between. That defines one fuzzy set, the set of those satisfied with their jobs, the happily employed. Now hands down. How many of you are not satisfied with your jobs? Many of the same hands go up again and bob and come to rest with elbows bent. This defines another fuzzy set, the unhappily employed, the opposite or negation of the first fuzzy set. A AND not-A. Now to some degree the Buddha's law holds. *Fuzzy logic is reasoning with fuzzy sets.*

The job sets differ from the male-female sets. The set of males does not intersect the set of females. No one is both male and female (in most audiences). Everyone is either male or female: A OR not-A. But most people are both satisfied and not satisfied with their jobs: A AND not-A. Few are 100% satisfied or 100% unsatisfied.

The audience example shows the essence of fuzziness: fuzzy things resemble fuzzy nonthings. A resembles not-A. Fuzzy things have vague boundaries with their opposites, with nonthings. The more a thing resembles its opposite, the fuzzier it is. In the fuzziest case the thing equals its opposite: the glass of water half empty and half full, the liar from Crete who says all Cretans lie and who both lies and not-lies, the borderline customer as satisfied as she is not satisfied. Here yin equals or balances yang, as in the ancient Taoist symbol (Figure 1.1).

The yin-yang symbol is the emblem of fuzziness. It stands for a world of opposites, a world we often associate with Eastern mysticism. The yin-yang symbol adorns the flag of South Korea. In Southern California it signifies a surf club.

FIGURE 1.1

For much of my youth I wrestled with this apparent mysticism in a world that science had whitewashed and blackwashed. Scientists had rounded off gray things to white and black things and then forgot about the rounding off and saw only a world of whites and blacks. The world is so much simpler if you can always cut the universe in exactly two pieces—if A OR not-A always holds. Modern scientists and philosophers would put a 1 or a 0, TRUE OR FALSE, next to every sentence in this book, instead of a truth fraction somewhere in between. The men and women of science beg the question of bivalence, assume the point at issue, climb the ladder of bivalence and forget they stand on it. The practice looks far more like religion than science. They turn their assumption of bivalence into an entrance exam and fail those who dissent, and they banish them with all the intimidation modern science can marshal: sloppy reasoning, not rigorous enough, unscientific measurement, untrained eye, poor experimental design, won't fit in a computer, commonsensical, folk psychology, would know better if you knew more math.

I had lost my faith in establishment science and found myself in a type of reverse atheism. I had passed the bivalent entrance exams but in my heart and head still failed them. I learned how to apply the rules of science but did not believe they were true. I learned how to manipulate probability but did not believe it existed.

Most of all the black-and-white world of science struck me as *unreasonable*, as when a zealous prosecutor or judge applies the letter and not the spirit of the law and you end up in jail if you spit on the sidewalk or deduct a nonbusiness dinner on your tax forms or mail-order the wrong magazine. Language, especially the math language of science, creates artificial boundaries between black and white. Reason or common sense smoothes them out. Reason works with grays.

I looked for an alternative that could challenge bivalent science on its own terms. If science rests on math, so should the alternative. Criticism fails without a working alternative. Fuzzy logic provided that alternative. It had the same math flavor that probability had, it worked with percentages between 0% and 100%, but it described events happening to some degree, not whether "random" events happened all or none. You paint one picture of the world if you say there is a 50% chance that an apple sits in the refrigerator. You paint a different picture if you say half an apple sits in the refrigerator. Same number, different worlds.

I pursued fuzziness through the channels of science and academia. I read and wrote papers on fuzziness, gave lectures and taught on it and videotaped new courses and seminars on it, sneaked it into the probability courses I taught at USC, helped organize conferences on it in the United States and Japan, wrote a textbook on it. I had to know if fuzziness existed. I was like a theist who had to know if God exists. If so, I would be a priest. If not, I would join the ranks of the opposition as an atheist zealot.

I looked for fuzziness and found it in a family of new math theorems, all housed in the geometry of a Rubik's cube, as we shall discuss. The math was so easy I could not believe someone else, everyone else, had not seen it. But soon I saw why even earlier fuzzy theorists might overlook this species of math or why they might see it as simple error if they looked at it at all. It involved strange notions like the whole contained in the part, big things stuck inside smaller things. I could understand why Western scientists did not want to turn fuzziness loose in the pristine world of black-and-white absolutes that they had built up over centuries. I could understand their fear of contradiction, their manic reactions to thing and nonthing, to A AND not-A.

I could understand the cultural prejudice and emotional reactions, all dressed up in the technical language and mannerisms of science, but I could not forgive them. Fuzziness solved age-old paradoxes of Western thought and opened new doors through mathematical infinity as it reduced black-and-white math to a special case of gray.

Most of all fuzziness made machines smarter. It increased the *machine IQ* of dozens of products in consumer electronics and manufacturing: cameras, camcorders, TVs, microwave ovens, washing machines, vacuum sweepers, transmissions, engine control, subway control. But it increased machine IQ in the land of A AND not-A, in the Far East, in Japan, where in the early 1990s fuzzy logic took hold inside and on TV sets, with even newscasters and politicians debating the meaning of fuzziness. Surely, I thought, money talks to scientists, since money drives everything in science and academia. But Western scientists and engineers only threw stones and we-could-do-it-toos at news of the fuzzy commercial successes in Japan. Earlier they had attacked fuzzy theory as lacking applications. Now they attacked the applications as lacking theory.

While Western scientists and engineers ignored or attacked fuzzy logic, their Eastern counterparts eagerly applied it and launched the long-awaited era of commercial machine intelligence. I often found myself reviled by Western scientists, especially the senior ones, including those in my own engineering department at USC—the grayer the hair, the more the reasoning seemed to be black and white. But in Japan I signed autographs and chaired conferences and waved at TV cameras. By the time we fuzzy theorists threw the first U.S. fuzzy conference in Austin, Texas, in June 1991 (at MCC or the Microelectronics and Computer Technology Corporation), the Japanese had already passed the $1 billion mark in annual sales of fuzzy products and taken another leap forward in their world leadership of consumer electronics and high-tech engineering and manufacturing. Cultural preferences come with costs.

Here is an IQ test for skeptics of fuzzy ideas: Can you explain how a fuzzy chip works? In the world of technology that question puts people into two nonfuzzy classes, the knows and the know-nots. Information does not respect rank or possessions or wrinkles. Logic helps people manage that information. Fuzzy logic helps machines manage it.

I don't think any human brain works with Aristotle's syllogisms or with computer precision. It's messier than that. The days of symbolic reasoning in "artificial intelligence" computer programs are over. They got unplugged with Hal the computer in the 1968 movie *2001: A Space Odyssey*. When Arnold Schwarzenegger's cyborg in *Terminator II* tells us it can learn new behavior because "My CPU is a neural-net processor, a learning computer," it does not mean Artistotle in a box. As we shall see, Aristotle ends up not in a box but at the corners of a fuzzy-logic cube, the rare moments of black and white in a world of gray.

If our reasoning has logic, it's fuzzy at best. We have only one decision rule: *I'll do it if it feels right*. The formal logic we first learn in tenth-grade geometry class has little to do with it. That's why we made it to tenth grade.

Fuzzy logic begins where Western logic ends.

2

THE FUZZY PRINCIPLE

Everything is a matter of degree.

ANONYMOUS

Shuzan (926–992 A.D.) once held up his bamboo stick to an assembly of his disciples and declared: "Call this a stick and you assert. Call it not a stick and you negate. Now, do not assert or negate, and what would you call it? Speak! Speak!" One of the disciples came out of the ranks, took the stick away from the master, and breaking it in two, exclaimed, "What is this?"

DAISETZ TEITARO SUZUKI
AN INTRODUCTION TO ZEN BUDDHISM

The fuzzy principle states that *everything is a matter of degree*. This book looks at the fuzzy principle in matters human, at how fuzziness permeates our world and our world views. We do not argue for the fuzzy principle so much as we look for it in whatever we look at.

Some things are not fuzzy no matter how closely you look at them. These things tend to come from the world of math. Here by design man or God has kept fuzziness out of the picture. We agree that "Two plus two equals four" is 100% true. But when we move out of the artificial world of math, fuzziness reigns. It blurs borders and deadlines as if our words cut the universe into pieces with a blunt knife.

Fuzziness has a formal name in science: multivalence. The

opposite of fuzziness is bivalence or two-valuedness, two ways to answer each question, true or false, 1 or 0. Fuzziness means *multi*valence. It means three or more options, perhaps an infinite spectrum of options, instead of just two extremes. It means analog instead of binary, infinite shades of gray between black and white. It means all that the trial lawyer or judge tries to rule out when she says, "Answer just yes or no."

Logicians in the 1920s and 1930s first worked out *multi*valued logic to deal with Heisenberg's uncertainty principle in quantum mechanics, which we will look at in a later chapter. This math principle says that if you measure some things precisely, you cannot measure other things as precisely. The principle suggests that we really deal with *three*-valued logic: statements that are true, false, or indeterminate. In short order Polish logician Jan Lukasiewicz chopped the middle "indeterminate" ground into multiple pieces and came up with many-valued or multivalued logic.* Lukasiewicz then made the next step and let indeterminacy define a continuum, a spectrum between falsehood and truth, between 0 and 1. On this "fuzzy" logic statements like "Grass is green" or "Lawyers settle disputes" can have any "truth value" or degree or fraction between 0 and 1, any percentage between 0% true and 100% true. The term "fuzzy" entered the scientific vocabulary about 30 years later. Until then logicians like Bertrand Russell used the term "vagueness" to describe multivalence. In 1937 quantum philosopher Max Black published a paper on vague sets or what we now call fuzzy sets. The worlds of science and philosophy ignored Black's paper. Else we might now be discussing the history of vague logic, not fuzzy logic.

In 1965 Lotfi Zadeh, then chair of UC Berkeley's electrical engineering department and later one of my Ph.D. advisers, published a paper called "Fuzzy Sets." The paper applied Lukasiewicz's multivalued logic to sets or groups of objects. Zadeh put the label "fuzzy" on these vague or multivalued sets—sets whose elements belong to it to different degrees, like the set of people satisfied with their jobs—to distance the concept from the runaway binary logic of his day. Zadeh saw

*Lukasiewicz had also studied Aristotle in the original Greek and worked out an early version of multivalued logic in the early 1920s.

scientists throwing ever more math at problems and trying to think and run the business of science with the black-white reasoning that computers and adding machines used. He chose the word "fuzzy" to spit in the eye of modern science.

The term "fuzzy" invited the wrath of science and received it. It forced the new field to grow up with all the problems of a "boy named Sue." Government agencies gave no grants for fuzzy research. Few journals or conferences accepted fuzzy papers. Academic departments did not promote faculty who did fuzzy research, at least who did only fuzzy research. The fuzzy movement in those days was a small cult and it went underground. It grew and matured without the usual support of subsidized science. That made it stronger.

Fuzzy logic did not come of age at universities. It came of age in the commercial market and leapfrogged the philosophical objections of Western scientists.

The fuzzy principle has emerged from almost three thousand years of Western culture, from three thousand years of attempts to deny it, ignore it, disprove it, relabel it, and axiomatize it out of existence. But fuzziness remains despite our best efforts to get rid of it. Our reasoning remains fuzzy. Now we can make our machines smarter by fuzzing up how they reason and fuzzing up the concepts, like *cool* temperature or *slow* speed, that they reason with. This creeping fuzziness confuses or angers many scientists who think machines must reason with black-white symbolic logic and math. This has led to a new joke in the world of artificial intelligence, where computer scientists still try to program computers with black-white logic to make them smart. The good news is at last the field of machine intelligence has made a product that is a commercial success. The bad news, at least to computer scientists, is it is a fuzzy camera.

In this book we will look at the fuzzy principle as it ranges from ancient Greece and India to smart washing machines in Japan to smart weapons in the future—and as it has crossed my path in science and engineering. I freely weave my fuzzy experiences into the discussion of the fuzzy matter at hand.

The book has three large sections: The Fuzzy Past, The Fuzzy Present, and The Fuzzy Future. The fuzzy-past section looks at the historical roots of fuzziness and antifuzziness, starting with how the world of Aristotle differs from the world of the Buddha, and reviews the nature of truth, Heisenberg's uncertainty principle of quantum mechanics, and the logical paradoxes of ancient Greece and modern math that led to fuzziness. The fuzzy-present section looks at fuzzy sets and systems and their recent rebirth in the United States and Japan and at how they extend to adaptive fuzzy systems, in which neural nets or "brainlike" systems let fuzzy systems learn from experience, let them grow their own rules from examples. The fuzzy-future section looks at how fuzzy logic and higher machine IQs may affect society in the near and distant future.

Several themes emerge from the journey from the fuzzy past to the fuzzy future. Four themes stand out to the greatest degree and we now preview them:

1. Bivalence vs. Multivalence: Simplicity vs. Accuracy
2. Precision Up, Fuzz Up
3. Fuzzy Reasoning Raises Machine IQ
4. Don't Confuse Science with Scientists

The first theme is the most important. It states the central conflict of the fuzzy experience.

THEME 1: BIVALENCE VS. MULTIVALENCE: SIMPLICITY *VS.* ACCURACY

Bivalence trades accuracy for simplicity. Binary outcomes of yes and no, white and black, true and false simplify math and computer processing. You can work with strings of 0s and 1s more easily than you can work with fractions. But bivalence requires some force fitting and rounding off, as when you state whether you are for or against the politician or satisfied or not satisfied with your job.

The Information Age rests on bivalence because it rests on the "digital revolution" in signal processing and microprocessor computer chips. We measure quantities—sound, blood pressure, light intensity, voltages, temperatures, earthquakes—that change smoothly with time. But we must sample and quantize, or round off, these signals to fit them into a computer's binary mind of 1s and 0s. You can view a time signal as a curve that wiggles from left to right. The curve might stand for the rise of temperature in an afternoon or the rise of ozone in a decade. Digitization imposes a grid on the curve, as in Figure 2.1.

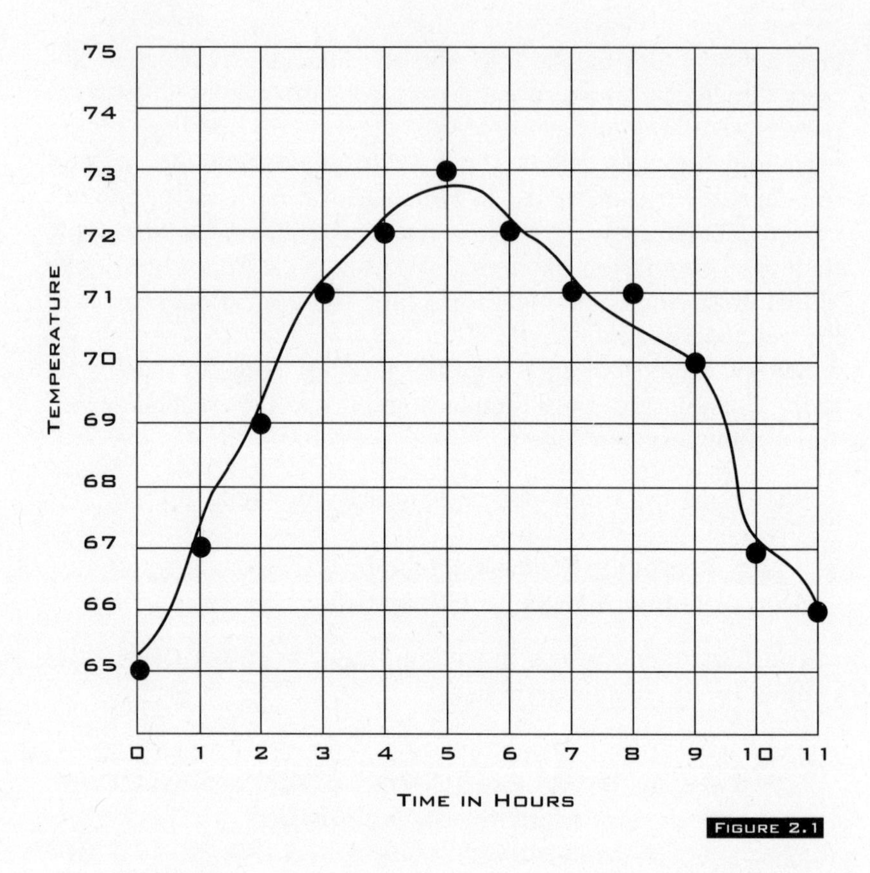

Figure 2.1

Digitization slices the bottom time line into discrete sample times—the times when a human or sensor measures an ozone sample and notes the time. Scientists have written thousands of journal papers about how to slice up the time line. Result: You should space the time slices evenly and the more slices the better. A compact disc plays a smooth sound curve built or converted from 44,100 samples per second.

Digitization slices the vertical line into a set of numbers. Here the system rounds off or "quantizes" the signal to the nearest sliced value. Then the system discards reality and keeps only the digitized numbers (the black dots on the grid) and converts each number into a unique list of 1s and 0s (32 = 100000, 35 = 100011). The rest is high-speed number crunching and a world of compact laser discs, cellular phones, fax machines, movie special effects, and new images of Neptune and Venus.

* * *

Western culture now sees binary precision as part of the scientific method. The digital revolution seems to have digitized our minds. Imagine a computer that answered "more or less" to a question. We might think the computer was slumming, programmed by a scientist in a white smock to talk to us as we talk to one another. We would not think that the computer meant it, that it spoke the truth, that the universe was that way, that somehow atoms and void added up to "more or less."

Aristotle's logic lies behind our bivalent instincts. We expect every "well-formed" statement to be true or false, not true more or less or false somewhat. A OR not-A. This "law of thought" runs through our language and teaching and thoughts. Existentialist philosopher Sören Kierkegaard titled his 1843 book on decision and free will *Either-Or* and saw man a cosmic slave to his binary choices to do or not do, be or not be. Today every engineering student takes a course in digital design and learns how to turn tables of binary logic into electrical circuits.

Every philosophy or religion has a villain or devil it seeks to avoid or destroy. The villain of bivalence is the logical contradiction: A AND not-A.

In bivalent logic a contradiction implies everything. It allows you to prove and disprove any statement. Mathematicians scour their axioms to keep them from implying statements that contradict one another. So far no one has proved that the axioms of modern math do not lead to statements that contradict one another. Tomorrow that may change and the framework of modern math may collapse. Meanwhile the dread and paranoia continue. So there is little tolerance in science for views that admit contradictions, that admit overlap between things and nonthings. Fuzzy logic confronts this intolerance head on. Fuzziness begins where contradictions begin, where A AND not-A holds to any degree.

Eastern mysticism offers the only major belief systems that accept contradictions, systems that work with A AND not-A, with yin and yang. Two hundred or so years before Aristotle lived, the Buddha would not let his audience trap him in either-or questions. He kept a "noble silence" when asked binary questions, such as whether the universe was finite or infinite. Modern Zen Buddhist monks train students to meditate on *koans*—what did your face look like before you were born? how does one hand clapping sound?—to break through the black-white shell of words and reach an enlightened or *satori* state of awareness. Even Chairman Mao Tse-tung wrote essays on contradictions.

The following list shows some of the key dual ideas and systems in human thought that this book looks at:

BIVALENCE	MULTIVALENCE
Aristotle	Buddha
A OR not-A	A AND not-A
exact	partial
all or none	some degree
0 or 1	continuum between 0 and 1
digital computer	neural network (brain)
Fortran	English (natural language)
bits	fits

In the last line *fits* stands for *fuzzy units*, just as "bits" contracts "binary units." A fit value is a degree or number between zero and one. A bit value is a zero or a one. A bit value answers a black-white question: Do you make over $30,000 a year? Do you own a car? Are you married? A fit value answers the same questions but only to some degree. (The fit value 70% means yes 70% *and* not-yes or no 30%.)

Consider again the apple you hold in your hand and bite. At first what you hold in your hand is 100% an apple. Or 100% of the apple is there. Or your apple belongs 100% to the set of whole apples. As you bite chunks out of the apple, the percentage falls from 100% all the way down to 0% when you have eaten the apple. About halfway through the process you hold the half apple or 50% apple. Fit values describe the apple's descent from total presence to total absence, from the bit value 1 down to the bit value 0. In this sense fit values "fill in" between bit values. If we graph how the apple falls from 100% apple to 0% apple, we see that the fit values fill in the number line between zero and one (Figure 2.2).

The number line shows the math conflict between bivalence and fuzziness. Bivalence holds only at the "corners" or ends of the number line. The bit values 0 and 1 stand as opposites.

Fuzziness or multivalence holds everywhere in between the

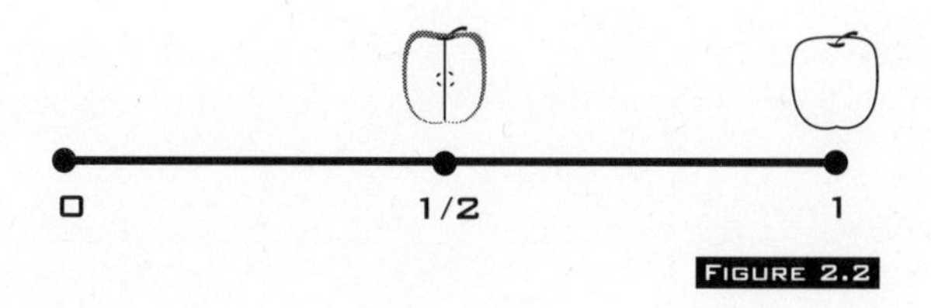

FIGURE 2.2

corners. It includes the two corners as special cases. When a glass has some water left in it and we round off and call the glass empty, we scrunch a fit into a bit, a 5% or 10% into a 0%. On the number line this means we first measure a fit value and then jump to the nearest corner of the number line to get a bit value.

You might here ask how we find the opposite of a fit value on the line. The opposite of 1 is 0 and of 0 is 1. Look at the line again. The opposites, A and not-A, reflect about the midpoint fit value of ½. The bit values 0 and 1 lie the same distance from the midpoint. The same holds for a fit value and its opposite. The opposite of ¾ is ¼. The opposite of ⅓ is ⅔ and so on. (This means that the opposite of ½ is ½. A equals not-A at the midpoint.) You can work with fits just as you work with bits. You don't have to round them off.

Rounding off fits to bits works well near the corners of the number line. But what happens if we want to round off the midpoint value? Do we round off 50% to 0% or to 100%? The question is not whether the glass is half empty or half full. If we had to say all or none, the question is, is the glass full or empty?

The midpoint of the line is a "paradox" in modern math. Mathematicians put the term "paradox" on the midpoint in part to suggest that midpoint or borderline cases are exceptions, simple problems we can fix with work. In fact they arise at the very foundations of bivalent math and logic.

Early in the twentieth century logician Bertrand Russell, who wrote the first peace sign, showed that paradoxes plague set theory, the mathematical theory of sets of objects and subsets of objects. He showed this with his barber example.

Russell's barber is a bewhiskered man who posts a sign that reads "I shave all, and only, those men who do not shave themselves." So who shaves the barber? If he shaves himself, then by his sign he does not. But if he does not shave himself, then by his sign he does. He seems to both shave and not shave himself at the same time.

Or consider the California bumper sticker that reads TRUST ME. We may trust the driver or not, or to some degree. But suppose we come upon a car with the bumper sticker DON'T TRUST ME. Do we trust the driver? If we trust the driver, then by instruction we don't. If we don't, then again by instruction we have trusted the driver. We end up both trusting *and* not trusting the driver, not an Aristotelian state of affairs.

The fuzzy interpretation sees the glass, barber, and driver as midpoint phenomena. Statements that describe them are literal half-truths. They are true 50%, not 100% or 0%. If we insist on a 100% full glass or a 100% shave or 100% trust, we land in bivalent paradox. The half-empty/half-full glass shows this. The water is 50% in the glass. That is a real state of the world. We don't mean that there is a 50% probability that the glass is full. We mean half a glass. If for some cultural reason we limit what we say to the two bivalent options of all or none, true or false, yes or no, then we pay the price and have a real contradiction on our hands, a case of A AND not-A.

Underneath the tug of war between bivalence and multivalence lies an equation. Bivalence says the equation does not exist or does not make logical sense. Multivalence says it exists to some degree. In extreme cases it exists to full degree or not at all. Editors prune equations from popular-science books as gardeners pull weeds from rose gardens. So to soften the blow of this equation, the central equation of the book and of fuzzy logic, I will give it a name that scientists and mathematicians are sure to ridicule, the *yin-yang equation*:

$$A = \text{not-}A$$

This is a "contradiction" in equation form. Instead of writing "A AND not-A" or "A is not-A" the equals sign equates the two propositions with all the rigor and pomp of formal math. In logic this means biconditionality: A implies not-A, and not-A implies A. So the paradoxes of bivalent reasoning reduce to the yin-yang equation: the half-empty cup implies that the cup is half-full and vice versa.

We can draw a picture of the yin-yang equation in action, or rather a sequence of pictures of where the yin-yang equation holds to different degrees. Recall the Venn diagrams of set theory taught in grade school. We cut a rectangle or box into two pieces, the A piece and the not-A piece, the A set and the not-A set. We split the box of apples into red apples and not-red apples. This gives a clean slice between the two sets A and not-A (Figure 2.3).

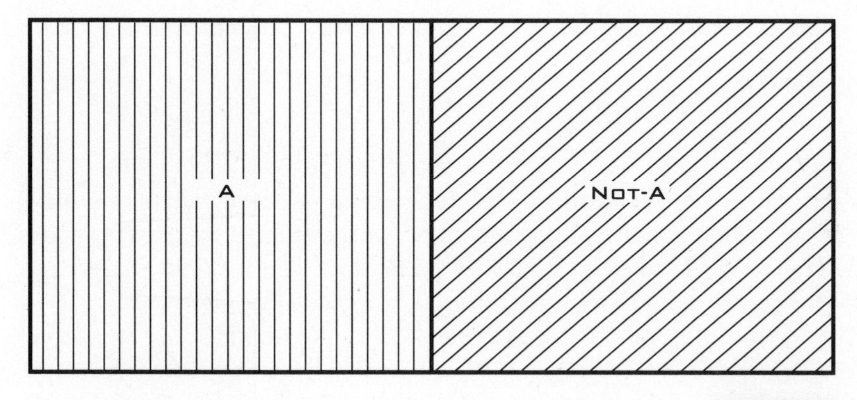

FIGURE 2.3

This is the bivalent case, the black-white world of math and Aristotle. The yin-yang equation does not hold at all here. More to the point it holds to zero degree. There is no overlap and there are exact boundaries.

Now suppose some apples are not all red—some have orange or pink or green streaks. If we ask a grocer to unpack the box of apples into two piles, the red apples and the not-red apples, she might form two piles and a third pile of apples hard to call one way or the other. The apples in the third pile are red to some degree and not-red to some degree. Most grocers will have a third pile of "gray area" apples that they find hard to classify. These break Aristotle's law of either-or. In the Venn diagram in Figure 2.4 the sets A and not-A have some overlap. In this case the yin-yang equation holds to some degree. The amount A overlaps not-A gives a measure of that degree.

Now suppose the grocer unpacks a new box of apples. This time each apple is as not-red as it is red. We do not care how the

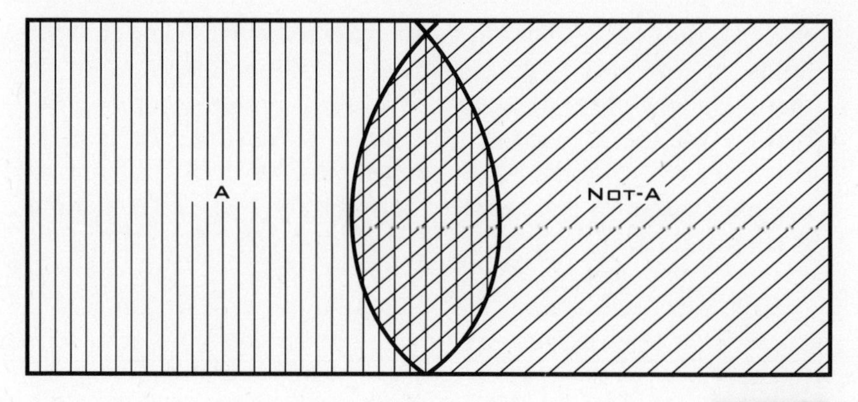

FIGURE 2.4

grocer measures redness and not-redness so long as she puts the apples into piles. Then the third pile becomes the only pile. There are no full red apples or not-red apples. So the Venn diagram shows two overlapping rectangles (Figure 2.5).

Here the opposites equal each other. So the yin-yang equation holds 100%: A = not-A. We cannot tell the thing A from its opposite not-A, the set of red apples from the set of nonred apples, the set of satisfied customers from those not satisfied, those who like their jobs from those who don't like them.

These three fuzzy Venn diagrams show once again that black and white are special cases of gray, that multivalence reduces to

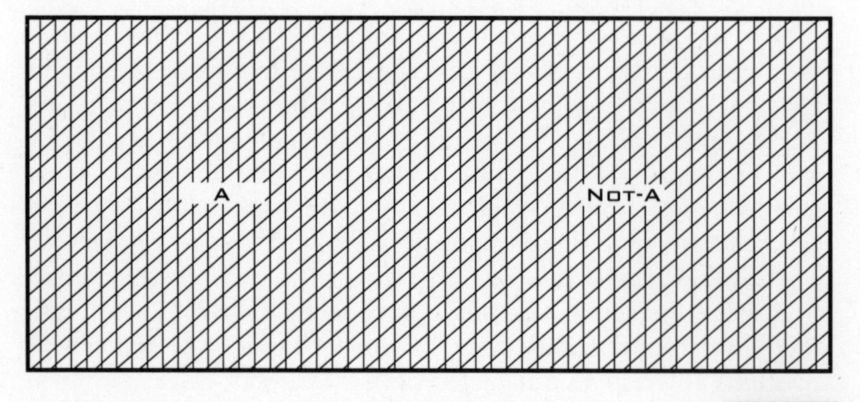

FIGURE 2.5

bivalence in extreme cases. In life as in Venn diagrams, we trade the expressive power and accuracy of fuzziness for the rounded-off simplicity of bivalence.

While teaching fuzziness in the mid-1980s I looked for a picture that captured the tradeoffs between fuzziness and bivalence. I found that picture in a Rubik's cube. The colored little squares were not part of that picture, though they are part of a more complicated picture of learning fuzzy systems, which we shall discuss in a later chapter. A Rubik's cube looks like a three-dimensional cube—a three-dimensional *fuzzy* cube. Any one of the six faces of the Rubik's cube looks like a two-dimensional cube or a solid square—a two-dimensional fuzzy cube. Any one of the twelve edges of the Rubik's cube looks like a one-dimensional cube or a straight line, the number line [0, 1] above—a one-dimensional fuzzy cube. You can see this with an apple example.

Look at three red apples. How many subsets are there of the three apples? There are eight subsets: all three apples, three pairs of twos, three pairs of ones, and no apples or the "empty set." A Rubik's cube also has eight corners or vertices. Coincidence? No, connection. In math we call it an "isomorphism"—the two objects have the same structure or morphology.

You can think of any three perpendicular edges of the cube as measures of length, width, and height. Or you can think of the three edges as measures of the redness of the three red apples. Each perpendicular edge ranges from 0% to 100% redness, from 0 to 1, from nonred to pure red. If we limit redness to the bivalent extremes of 0 or 1, that gives eight combinations of 0s and 1s at the corners of the cube (Figure 2.6).

The 1s and 0s stand for the total redness or nonredness of an apple. The binary string (1 0 1) means the first and third apples are 100% red but the second apple is 100% nonred. The zero binary string (0 0 0) means no apples are red, the empty set of red apples. If you look at the cube for a while you can see Aristotle's laws of black-white logic at work. Long diagonals (the dashed line) connect opposites. In the figure a long diagonal connects (0 0 1), the set that contains the third red apple, with (1 1 0), the set that contains only the first and second red

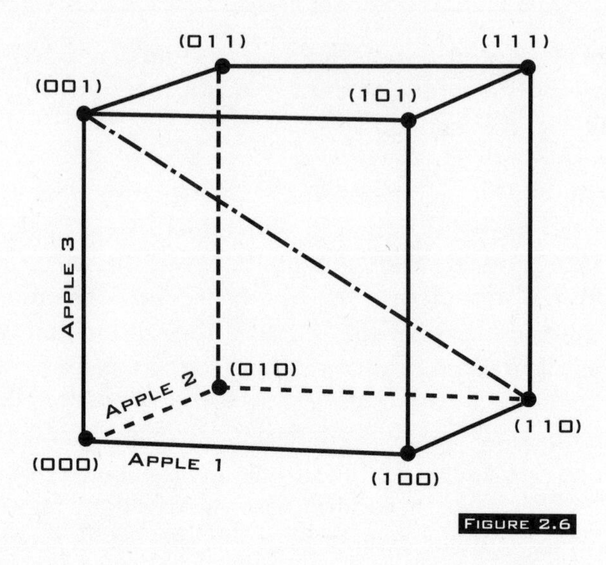

FIGURE 2.6

apples. You get bivalent opposites if you replace 1s with 0s and vice versa.*

If bivalent sets lie at the corners of a cube, what lies inside? Fuzzy or multivalued sets. In these sets the apples are red only to some degree between 0 and 1. The fuzzy sets fill in the cube. The fuzzy set (0 0 ¾) means that only the third red apple is present and it is only 75% red. The set of half-red apples (½ ½ ½) lies at the midpoint of the fuzzy cube. This set not only equals its own opposite—the set of half-unred apples—but it, and it alone, lies the same distance from all eight corners of the cube. You cannot round off the midpoint to the nearest corner, because every corner is equally near and equally far.

---

*The law of excluded middle (A OR not-A) holds because if we lay the bit lists (0 0 1) and (1 1 0) side by side and pick the largest bit value in the corresponding slots we get the union bit list (1 1 1):

(0 0 1) or (1 1 0) = (1 1 1)

So every red apple is in the first set or the second set of apples. This union bit list also arises if we add corresponding bit values. The law of noncontradiction not-(A AND not-A) holds because if we pick the smallest bit value in the corresponding slots in (0 0 1) and (1 1 0) we get the empty set (0 0 0):

(0 0 1) and (1 1 0) = (0 0 0)

So no apple is in both the first and second sets of apples. This intersection bit list also arises if we multiply corresponding bit values.

Consider an audience of two people. Each raises one hand to answer a question. We ask them if they are happy with their jobs. In the extreme case of Aristotle's black-white logic, four things can happen: (1) both hands go all the way up, (2) both hands stay down, (3) the first hand goes up and the second stays down, or (4) the first hand stays down and the second goes up. These four bivalent cases correspond to the four pairs of bit values (1 1), (0 0), (1 0), and (0 1). The 3-D fuzzy cube has lost a dimension and become a 2-D cube, a unit square (Figure 2.7).

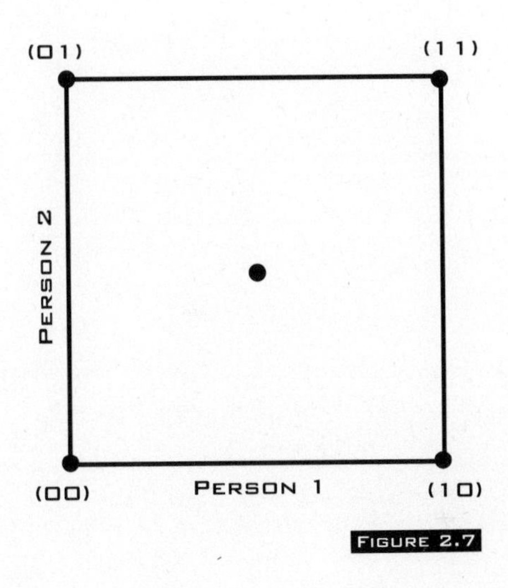

FIGURE 2.7

The horizontal axis now measures the degree to which the first person raises her hand. The vertical axis measures the degree to which the second person raises his hand. The midpoint stays in the middle, still the same distance from each of the four bivalent corners.

Fuzzy audience responses fill up the rest of the square. The fuzzy response (⅓ ¾) is a list of fits (fuzzy units) instead of bits. Here the first person raises her hand only partially, about 33.3%. She dislikes her job more than she likes it. The second person raises his hand 75%. He likes his job more than he dislikes it. To place them in the fuzzy cube we start at the lower left corner and slide over to ⅓ on the horizontal line and slide up to ¾ on the vertical line (Figure 2.8).

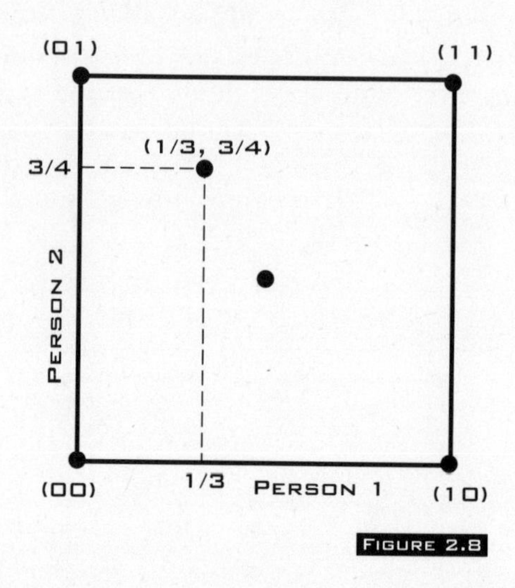

FIGURE 2.8

The opposite audience response is (⅔ ¼). Here the woman likes her job more than she does not like it. The man dislikes his job more than he likes it. If the letter A stands for the first fuzzy set (⅓ ¾), then not-A stands for its opposite (⅔ ¼). Now what happens to Aristotle's laws? The either-or law no longer holds: A OR not-A equals (⅔ ¾). It falls short of the bivalent extreme (1 1). The law of noncontradiction also fails: A AND not-A equals (⅓ ¼) and not (0 0). It is not the "empty set" (0 0). Aristotle has lost some ground to the Buddha—when one goes up the other goes down. If we plot these four points inside the fuzzy cube, we see that they define an interior square (Figure 2.9).

I call this "completing the fuzzy square." As the audience responses get less fuzzy, the interior square moves out toward the nonfuzzy corners. In the extreme case, if we follow the bivalent method and round off the fuzzy responses (⅓ ¾) to the nearest bit values, we get (0 1), the upper lefthand corner. In that case the interior square equals the corners of the cube and Aristotle reigns. Modern science and math stand on exactly this

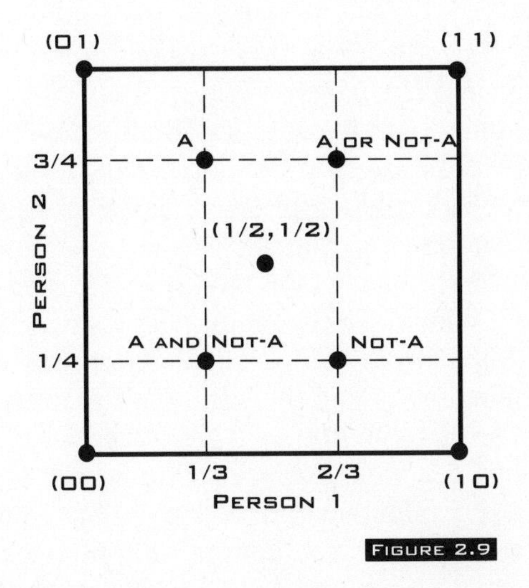

FIGURE 2.9

rare case of all answers black or white—Aristotle boxed into corners.

Now let the answers get fuzzier. Say they tend to raise their hands more or less than halfway up. Then the interior square of A, not-A, A OR not-A, and A AND not-A shrinks in to the midpoint. If the audience has fifty or a hundred persons in it, the same thing happens but in "hypercubes" of high dimensions that we cannot visualize.*

The fuzzier the answers, the more A OR not-A looks like A AND not-A, the more the interior square shrinks. In the extreme case when every hand rises only 50% and each person is as happy with his job as he is not happy, the interior square collapses into the cube midpoint. Then, and only then, does the yin-yang equation hold 100% inside the cube. Then the thing A equals its own opposite not-A:

A = A OR not-A = A AND not-A = not-A

At the midpoint you cannot tell a thing from its opposite, just as you cannot tell a half-empty glass from a half-full glass. At the

*The fifty-person cube has 2^{50} corners, where 2^{50} means $2 \times 2 \times 2 \times \ldots \times 2$, 50 times. The hundred-person cube has 2^{100} corners. All cubes have exactly one midpoint. It alone among all points in the cube lies the same distance from each corner.

cube corners you can tell a thing from a nonthing with 100% clarity. In between lie shades of gray.

Bivalence holds at cube corners. Multivalence holds everywhere else. Aristotle rules at the corners of the fuzzy cube, the rare cases of black-white options amid a continuum of gray options. The Buddha rules at no corner. The Buddha rules to some degree everywhere inside the cube. He rules 100% at the cube midpoint, where the yin-yang equation holds 100%. You can picture this with a lot of small Aristotles sitting at the corners of a cube and a Buddha sitting in full lotus at the center of the cube. The cube midpoint is full of paradox. It sits like a raised middle finger to the black-white world of science. No matter how hard bivalent scientists try, they can never round off the midpoint to a corner. You cannot both fill and empty a half-full glass of water. The midpoint is the black hole of set theory.

THEME 2

Precision Up, Fuzz Up

Information helps us picture the world. Every second our eyes transmit millions of bits of information to our brains. Our minds feed on newspapers, TV shows, telephone calls, letters, faxes, and gossip. We expand our senses with microscopes, contact lenses, binoculars, thermometers, barometers, CAT scanners, telescopes, and hundreds of other devices that help us convert the world into information.

Each new datum changes our mind. Inside our brain it changes how our brain cells, or neurons, fire. This changes slightly the patterns our synapses, the "wet" wires between our brain cells, learn or remember. As we gain more information, we get a clearer, more accurate picture of the world. We get a clearer view of facts. But does that take the fuzziness out of the facts?

Suppose John is in his early thirties. Is John old? Yes or no? Is he young? Yes or no? Add some information. Pin down John's age more precisely. Say we know John is age 30 to the day (it's his birthday). So is John old or young? What does the precise information about John's age tell us? It tells us only that John will be older when he turns 35 than he is now. That tells us that OLD and YOUNG are matters of degree. They are fuzzy concepts. OLD and YOUNG stand for fuzzy subsets of the human population.

It all comes down to this: Where do we draw the line? That question haunts black-white reasoning in a world of grays. The U.S. Government says adulthood begins the first second of your eighteenth birthday. The government draws the line for us and we make the best of it. We can view this line as the imaginary line that splits ADULT from NONADULT on the scale of years (Figure 2.10).

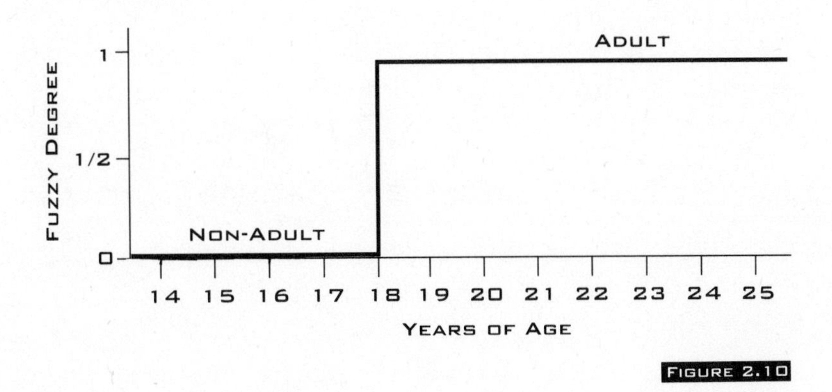

FIGURE 2.10

We can draw different lines at different ages near age 18 but we cannot give a good reason for them. We know that the 14-year-old is barely if at all an adult and the 25-year-old is usually if not fully an adult. We also know, as with the fuzzy concept OLD, that adulthood grows with age. So the fuzzy principle views ADULT as a fuzzy concept and draws it as a curve, not a line (Figure 2.11).

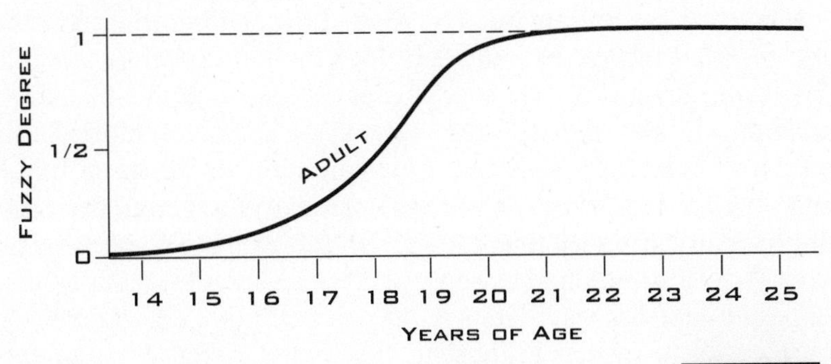

FIGURE 2.11

Fuzzy theory *draws a curve* between opposites, between A and not-A. More information, more "facts," help us draw the curve. If we have enough information, we can turn our vague notions of OLD and YOUNG into fuzzy-set curves (Figure 2.12).

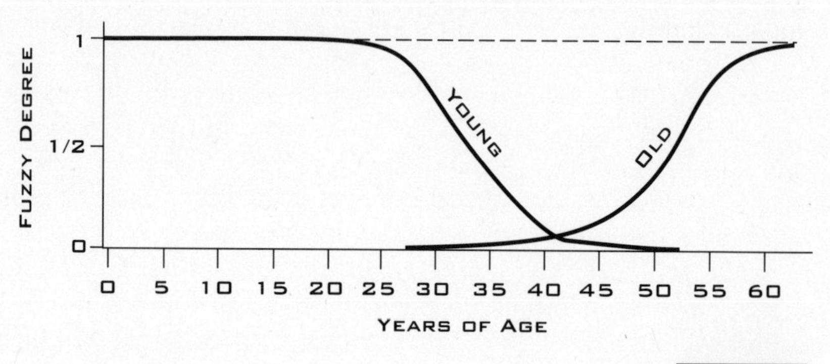

FIGURE 2.12

The more information we have, the more bumps in the curve, the more realistic the curve.

The split between thing and nonthing grows more complicated still when the thing is relative. Consider the "line" between art and nonart. Does more information help a censor decide what paintings, body movements, lyrics, film sequences are art and which are "obscene" or nonart? In the end the split is a "judgment call." And in the beginning it is too. Objects and behavior are to some degree art or nonart depending on taste, tradition, and whim. Art dealers and critics rank and price paintings, drawings, and sculptures according to what they see as the objects' *beauty* content.

Beauty is both fuzzy and relative. It depends on the speaker and on the culture. Similar paintings or melodies or story plot lines and phrases can start a cultural trend or end in a plagiarism conviction. Pop artists and comedians excel at blurring the borders between beauty and entertainment or between insight and absurdity or between art and a censor's unstated definition of obscenity. Beauty lies not only in the eye of the beholder, it lies there to some degree.

Legal decisions are also fuzzy and relative. The scales of justice tip to varying degrees. Courts convict persons who commit crimes with enough intent and acquit those who commit them with enough diminished capacity. Judges, legal scholars, and the rest of us search for the borders between personal freedom and government control, between man and state, choice and command. Jokes shade into insults or slander or harassment. You may own the land your house rests on, but do you own the air space above your house? Do you own the dozens of radio and television signals that right now propagate through your body? Who owns the oceans or the Moon or the Sun or the Oort Cloud? What if we dig up ancient property markers, millions or hundreds of millions of years old, from spacefaring aliens who have just sold the planet to another race or who seeded life here and plan to harvest it?

Legal concepts vary among cultures and within them. The great increases in information in the twentieth century have not helped us draw lines between justice and injustice, fair and unfair, right and wrong, intention and no intention, contract breach and nonbreach, private and public, mine and thine. Information will increase for centuries. Rather than simplify legal decisions, more information increases the fuzz in and between legal thing and nonthing. It deepens the legal quagmire. Precision up, information up. Information up, fuzz up.

Is life a fuzzy concept? Will more science and measurements resolve the abortion debate? Will medical science or the state draw the line between life and nonlife at conception or at the first or second trimester? In time I think we will draw life as a fuzzy curve (Figure 2.13).

We may draw this curve with data from medical conferences or

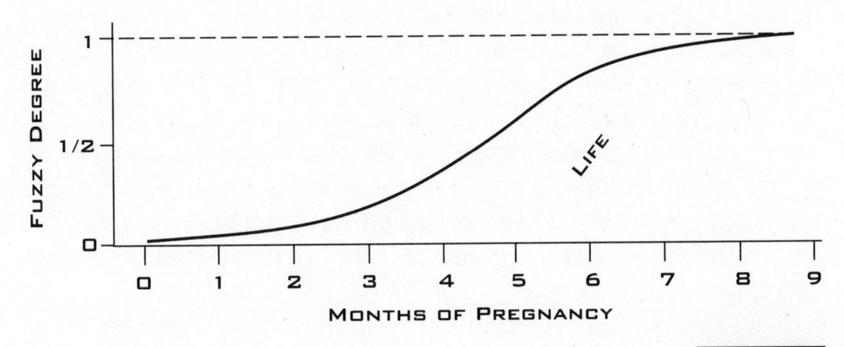

FIGURE 2.13

media polls or national elections or interactive cable plebiscites. What we do with such a curve is another matter—another matter of degree.

More information means more facts. More information will better describe the facts. It will give us clearer pictures of facts from more angles. But fuzz promises to be a permanent part of those pictures. In many ways the future looks fuzzy.

THEME 3:
FUZZY REASONING RAISES MACHINE IQ

Fuzzy engineers design software and chips to make computers reason more as people do. That makes them smarter and easier to work with. Now you can program fuzzy software systems in English or Japanese and leave the details to software engineers you may never see. Future fuzzy systems will allow you to program them by spoken language instead of stroked keyboard. Other future fuzzy systems, the adaptive fuzzy systems we discuss in a later chapter, learn from experience and program themselves.

Fuzzy knowledge comes down to fuzzy rules. A fuzzy rule relates fuzzy concepts in the form of a conditional statement: If X is A, then Y is B. If the traffic is HEAVY, then keep the traffic light green LONGER. Just common sense. The rule associates the fuzzy sets HEAVY and LONGER. We can draw these sets as curves or as points in big fuzzy cubes. All traffic is HEAVY to some degree. We can make the time the light stays green LONGER to some degree. The combinations of traffic densities and light times are infinite. Yet one fuzzy rule connects them all—and raises the machine IQ of the traffic controller that uses it.

Fuzzy systems store dozens or hundreds or thousands of these common-sense fuzzy rules. Each new piece of data activates all the fuzzy rules to some degree (most to zero degree). The fuzzy system then blends together the outputs and produces a final output or answer. On a fuzzy chip this "parallel" reasoning takes places thousands or millions of times per second. We count the reasoning in FLIPS or fuzzy logical inferences per second.

High-speed fuzzy systems are smart. Today in Japan they control subways and stabilize helicopters better than humans can. The fuzziness in their rules leads to smooth control. This cuts down on the jerky overshoot and undershoot of the old math-control systems: the air conditioner that blows too cool or too warm, the camera that underfocuses and blurs or overfocuses and blurs. Soon we will have fuzzy devices in our homes, offices, cars, and aircraft. We may not know it, and the advertisers may not tell us, but they will be there. We will command armies of tiny high-speed fuzzy experts that never err or tire or complain.

Sensor technology speeds the fuzzy revolution. Those tiny fuzzy experts need lots of data and the faster and the more precise the better. A fuzzy washing machine uses load sensors to measure the size and texture of the wash load and uses a pulsing light sensor to measure the dirt in the wash water. Each second a few fuzzy rules turn these measurements into patterns of water agitation for different lengths of time. Fuzzy vacuum sweepers use infrared sensors to measure dirt density and carpet texture. The data comes in and the fuzzy rules adjust the sweeper's sucking power. Fuzzy TVs measure the relative brightness, contrast, and color in each TV image frame and then "turn the knobs" on these values for each part of each image to give a sharper picture. As you watch the TV the knobs keep turning slightly as if a highspeed expert watched the screen and worked out the best mix of settings for each of the 30 or so images that flash by per second.

In older fuzzy systems an expert gives the common-sense rules. A fuzzy engineer may sit down with an expert and ask her how she focuses a lens or makes a left turn or steadies a helicopter. In *adaptive* fuzzy systems a "brainlike" neural network, a computer system that mimics how brains learn and recognize patterns, generates the fuzzy rules from training data. They learn from experience. DIRO: Data in, rules out. The neural system behaves like the eyes and ears of the system. It "sees" patterns in the data and slowly grows rules that relate these patterns. The patterns are fuzzy sets and the relations are fuzzy rules. The fuzzy system uses these rules to reason with the patterns. You would show an adaptive fuzzy traffic controller examples of how a traffic cop or a traffic light controls traffic. More data in, better rules out.

Adaptive fuzzy systems "suck the brains" of experts. Experts do not have to tell the system what makes them experts. They just have to act as experts. That gives the data that the neural nets use to find and tune the rules. Superhigh machine-IQ systems of the future, whether in smart cars or in smart missiles or in tiny robots that swim in our blood and tissues and on our teeth, may grow all their fuzzy rules with neural nets. After all, we do.

THEME 4: DON'T CONFUSE SCIENCE WITH SCIENTISTS

Scientists have in large part treated fuzzy theory and fuzzy theorists badly. Some of us asked for it. All of us got it. In the end that process strengthens fuzzy theory and fuzzy theorists. Adversity, like muscle stress, works that way.

In the meantime many of us lost our faith in science. That was a deep disappointment for those of us who had earlier lost faith in religion and government. Science was not salvation. Career science, like career politics, depends as much on career maneuvering, posturing, and politics as it depends on research and the pursuit of truth. Few know that when they start the game of science. But they learn it soon enough.

The hardest things I learned in my fuzzy quest were that modern science does not welcome a truly new idea. And it makes mistakes even at the "self-evident" level of logic and math.

Science prefers small steps to large creative leaps. Modern science often behaves no better with new ideas than the Roman Catholic Church behaved when it forced Galileo to renounce his belief that the Earth rotates about the Sun. Unlike the church, modern bivalent science does not claim to possess all knowledge. It claims to follow the only road to knowledge. And by a cultural fluke—the flowering of ancient Greece—it worked black-white logic into that road. Science assumed bivalence as true and so saw fuzziness as not scientific. I once called a journal editor after he had rejected my fuzzy response to a probability paper. His letter had said his journal could not publish a paper that claimed A AND not-A is okay. So I called him and asked him why. He came right to the point: "Because I'm the editor."

My fuzzy work and fight also taught me a hard fact: science differs from scientists. The product of science is knowledge. The product of scientists is reputation.

Everyone pursues fame and power to some degree. Scientists are unique in their pursuit for at least two reasons. First, their product is reputation. Second, they answer to no higher authority. The institution of science provides scientists with the means to the personal ends of fame and power, and they train much of their youth to enter this institution. If all goes well, the scientists build strong reputations and leave behind them the product of science, knowledge. They leave journal papers, textbooks, monographs, conference proceedings, software, and even new hardware devices. They take home pay checks, professional awards, well-shook hands after speeches, speech honoraria, prize money, consulting checks, board-of-directors stock, and most of all they take home pride.

Fuzzy logic has added to science. It has extended the binary math that underlies science. It has helped solve the mismatch problem by allowing us to paint gray pictures of a gray world. As discussed in the next chapter, fuzziness has reduced the ancient and undefined concepts of "randomness" and "probability" to a natural relation that holds between sets. Fuzzy logic has added to machine intelligence by showing us how to make machines reason more as we reason, with common sense learned from experience, and has so raised machine IQs. In many cases fuzzy logic made these contributions at the expense of the collective reputation of tens of thousands of living and dead scientists reared on the bivalent faith. Since, according to futurists, 70% to 90% of all scientists and engineers who ever lived are alive today, most of the collective expense has fallen on modern scientists.

Last century John Stuart Mill said that new ideas pass through three phases of denial. First, they are wrong. Second, they are against religion. Third, they are old news, trivial, common sense, and we all would have thought of them if we had had the time, money, and interest. Fuzzy logic, tied to society's rapid rate of change of information, is passing from phase one to phase three. In the West fuzzy theory lies between phase one and phase two. Most scientists still attack it as against the bivalent faith. Only the commercial success of fuzzy products has made it an issue. Else it might still lie underground in articles in obscure journals. In the Far East fuzzy theory has advanced almost to phase three. There objections to fuzzy logic have more to do with technical than philosophical issues: the worth of expert knowledge, the availability of training data, ease of software development, computer-chip space or speed, accuracy of sensor data. At root fuzzy logic or multivalence is a world view or ideology. Bivalence is an ideology too, and that is where the conflict lies.

Scientific ideologies clash in a political arena. I know this subject is distasteful to my colleagues but it too often goes unsaid and someone must say it. The history of fuzzy logic is steeped in these politics. Outsiders may mistake the politics for scientific procedure—if they can cut through the technical jargon and thick writing. Man, as Aristotle said, is a political animal. The politics runs unchecked because both scientists and nonscientists have draped scientists with the mantle of truth seeker, the shield of specialized knowledge.

The press seldom penetrates this shield. When they do they do not know how to navigate through the technical subject matter and do not know whom to trust in a dispute. So the press defers to an unnamed "majority" of scientists and prints the opinions of a few and then abandons the topic. They do not deal so lightly with the politics of government, industry, sports and entertainment, or the press itself. How many *60 Minutes* programs have you seen on academic fraud and infighting, graduate-student abuse, or government appropriation and management of research funds?

Every day scientists make choices from large sets of alternatives. Politics lies behind literature citations and omissions, academic promotions, government appointments, contract and grant awards, conference addresses and conference committee-member choices, editorial-board selection for journals and book series, reviewer selection for technical papers and contract proposals and university accreditation status, and most of all, where the political currents funnel into a laserlike beam, in the peer-review process of technical journal articles—when the office door closes and the lone anonymous scientist reads and ranks his competitors' work. As they say at technical conferences in the poster sections, where those denied a formal oral talk show their results on a lonely poster board, publish or perish—help your friends publish and help your enemies perish.

To what end? A degree, a job, a promotion, tenure status, a raise, a grant, better consulting jobs, another listing on a résumé, an award, an appointment, a plenary lecture, name fame, posthumous fame, to testify before politicians, maybe to become a politician, to even scores, to help friends who help friends, the sheer fun of beating your competition. To the end of reputation.

Nuclear weapons brought together science and lawyers in the same office. The deal was clear in those early days: lawyers over scientists. Most of the scientists were physicists and economists. They raised new policy options for politicians. They did not dictate the policy options and few became career politicians.

Today that has changed. More scientists from more fields help the government face new problems and draw up new policy options. We live in an age of lawyers *and* scientists. Global warming. Global economics. Greenhouse emissions. Carcinogenic foods. Psychological trauma. Genetic engineering. Abortion. Space exploration. Wars on drugs. Drug legalization. Smart weapons. Life extension and cryonics. To some degree every item on the social agenda or on the nightly news deals with the opinions of scientists and engineers. Politicians fund and review scientific studies, call scientists to testify before them, and then draw up rules and laws. Journalists, one fuzzy step removed from elected politicians with their TV and print spotlights, cite the most recent "scientific" polls to support their views. Print and broadcast journalists help shape those public opinions.

The next step is scientists over lawyers. I do not look forward to that day, though as a scientist I might do well in such a science state, whenever it comes. My fuzzy past makes me wary: *if scientists can err at the "self-evident" level of logic and math, they can err at anything*. That is the real lesson of the fuzzy experience and we should shout it from rooftops. As the political-science proverb goes, I would rather be governed by the first thousand names in the phone book than by the faculty of MIT or Stanford or Tokyo University.

There is one more theme that runs through this book: the fight between probability and fuzziness. Scientists have long said that fuzziness is just probability in disguise and that what you can do with fuzzy logic you can do without it.

I spent part of my youth in this fight and it shaped much of my world view. That takes a chapter to tell and leads to the deepest and strangest idea in fuzzy logic: the whole in the part.

3

The Whole in the Part

The method of postulating [assuming] what we want has many advantages. They are the same as the advantages of theft over honest toil.

Bertrand Russell
Introduction to Mathematical Philosophy

In the beginner's mind there are many possibilities. In the expert's mind there are few.

Shunryu Suzuzi
Zen Mind, Beginner's Mind

For years I wanted to know how fuzziness related to probability, the mathematical and philosophical theory of "randomness" and luck. Is fuzziness just "probability in disguise"? What is probability? What is randomness? Can you define one without the other? What is chance?

One night in a hot tub I found an answer: the whole in the part. That answer comes at the end of a long hunt for probability. It also lays the foundation of fuzzy logic. To start the hunt consider this question: Can you draw a circle?

No one has ever seen a circle. No one has ever seen a square or a triangle or an ellipse or any other geometric object. We have seen only approximations, imperfect grays instead of perfect blacks and whites. Zoom in close enough and you will see imperfections in the drawing or printing or engraving or assembly of subatomic particles. Maybe God or superaliens can draw perfect circles and squares right down to the last quark and farther. And maybe they can't.

This relates to probability in a simple way. Try to find probability in the inexact circle or oval in Figure 3.1.

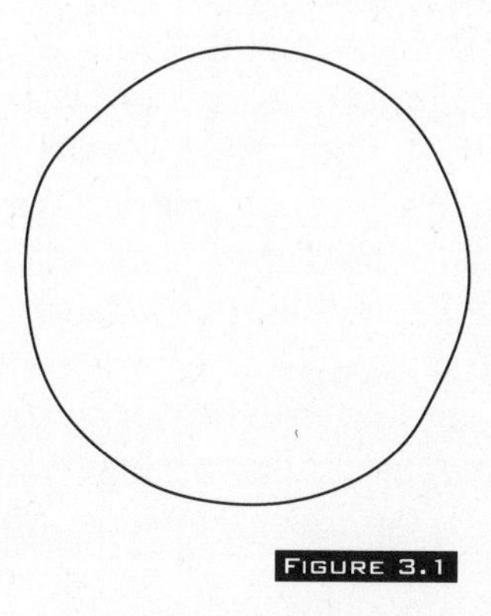

FIGURE 3.1

Now consider two competing views of the inexact oval. *Probability view*: The oval is probably a circle. *Fuzzy view*: The oval is a fuzzy circle. Which is more accurate?

The probability view commits to a property hard to find in the oval. Where is the "randomness"? The figure is fixed and static. It does not vibrate on the page. We can measure the inexact oval until all the facts are in and that will not pin down the "randomness." The more information we have about a fact the less we tend to blame the fact on probability or "luck." Total information leaves little room for probability.

We often use probability talk when we are not sure about something. He will probably buy a sports car. You are probably catching a cold. Napoleon probably thought the Prussians would not make it in time to Waterloo. You are probably right.

For centuries we have used probability talk to describe uncertain acts and events. You even find probability modifiers in the cuneiform text of ancient Sumeria and Babylon. Probability has become our cultural default for uncertainty, the one-size-fits-all description for anything we are not sure of. In casual talk, or in a scientific paper loaded with math, we might use the modal adverb "probably" to describe the oval figure's inexactness, how it deviates from circular perfection. What a strange adverb "probably" makes. It casts a "random" mood over our verbs and descriptions and thoughts. It takes us from a simple state of being in "That is so" to a new world of quivering chance in "That is probably so."

Now consider the fuzzy view of the inexact oval. The adjective "fuzzy" means that the figure is to some degree a circle and to some degree not a circle, but more a circle than not. It reminds us that in practice we cannot draw a hard line between circle things and noncircle things. They overlap. In that region circles shade into noncircles. The fuzzy view sees an ambiguity or vagueness between thing and nonthing, have and have not, circle and noncircle, A AND not-A. That differs from whether a perfect circle "happens" at random or whether, if trained in probability and statistics, we conduct a coin-tossing "random" experiment in our head and believe the inexact oval results as an outcome.

One day as a graduate student I drew an inexact oval and stared at it. I could find no probability or lucklike uncertainty in the figure. All the probability math I had learned in graduate school did no good. I could find no probability. I could not find what at first made me want to say that the inexact oval was *probably* a circle. And the inexact oval was no exception. I could find no probability in anything anywhere.

Yet probability seemed to be at work everywhere. Physicists said every thing and every subatomic particle and every particle interaction comes from a random experiment. They gave Max Born the Nobel Prize in physics for his claim that quantum matter waves are probability waves.* Erwin Schrödinger gave the equation that shows how matter waves change in time. Born saw the matter waves as the underside of probability. Matter waves burn as probability fire. The matter was not there, it was

---

*In quantum mechanics the psi operator ψ stands for the stationary matter wave in some infinitesimally small region or volume dV. Max Born interpreted the squared absolute value of the matter wave, $|\psi|^2$, as a probability measure. Then the infinitesimal quantity $|\psi|^2$ dV measures the infinitesimal probability that a particle of matter, say an electron, lies in the infinitesimally small region dV. This implies that all infinitesimal particles are random points. In contrast the fuzzy view sees $|\psi|^2$ dV as the *degree* to which the particle resides in the region dV. The particle resides in all regions of space to some degree. This means that particles are deterministic clouds.

probably there. Matter described its own probability. Einstein said everything was relative in terms of where you sit when you measure time or distance. Born said everything is probable and meant it in the sense that everything *is* probability. Science rewarded that radical random metaphysic with its biggest prize. I am glad Dr. Born got his prize but worry that it stopped dissent on the matter.

Back to the hunt. Where is the probability? Everywhere. Where is the randomness? Everywhere. Examples should be as plentiful as quarks and leptons and hydrogen atoms. But we find only the footprints of randomness. We find only the after-the-fact outcomes of random experiments. Everything is a footprint. We never catch probability in the act.

THE GOD OF MAXIMUM PROBABILITY

Scientists have refined the probability world view in the twentieth century. We do not observe just probability outcomes. We observe *maximum* probability outcomes, the most likely outcomes. Everything we observe is the outcome of complex physical, chemical, biological, social, and cosmological processes. Each process unfolds as an event chain. Each future event can go a million or trillion different ways. Science says the different ways have different probabilities attached to them and you should expect the most probable ways to happen.

This *maximum likelihood* view lies at the base of modern science and engineering and much of common sense: We "tend" to observe the most probable events. We say "possible not probable" about whether the long-shot candidate will win the election or whether it will snow in Los Angeles or whether aliens visited earth millions of years ago and genetically engineered ape-men from apes. We expect the most-probable outcomes to occur even if we silently buy a lottery ticket or play a slot machine or smoke a cigarette. We play the percentages.

Science plays the percentages too. It holds the maximum likelihood view as the grand organizing principle of the scientific world view. In the next second all the air molecules in the room may fall into a small dense air bubble in the center of the room and leave the rest of the room in vacuum and suffocate anyone in the room. That event could happen but has small probability.

Possible not probable. Such air-bubble molecule distributions are a minute fraction of the number of all possible air-molecule distributions in the room. The great majority of the molecule distributions are uniform throughout the room. "Thus" we observe and breathe uniform distributions of air molecules.

For the same reason most strings of text are not words. Say you make 100 copies of each alphabet letter and write each letter on a Ping-Pong ball and put all the Ping-Pong balls in a bingo hopper and mix them up. Now reach in the hopper and "at random" pull out the first six or eight or ten Ping-Pong balls. The odds are great that your string of six or eight or ten letters will not spell out a word in the dictionary.

The maximum probability view permeates science from the subatomic level to the cosmic. Our universe may not be, as Leibniz said, the best of all possible universes but it is the most probable. Leibniz was wrong. Statisticians were right.

Cosmologists like Stephen Hawking give a "maximum likelihood" argument for why our universe began in a Big Bang explosion that produced irregular clumps of matter or galaxies instead of an expanding uniform bubble of mass. There are many possible Big Bang "initial conditions" or starting points. In theory there are infinitely many initial conditions. Different initial conditions evolve into different universes just as different cherry seeds grow into different cherry trees. Very few initial conditions lead to uniform-mass universe bubbles. Since they are scarce they are "improbable." They are possible not probable. Most initial conditions are nonuniform. So they are "probable." Nonuniform initial conditions are highly asymmetric and unstructured and resemble knotty jagged portions of an ocean coastline. Our universe seems to have evolved from one of these choices of nonuniform initial conditions.

Who chose our initial condition? The God of Maximum Probability chose it. Someone or No One reached in the box of universe seeds and pulled out a plain old seed. And here we are, the luck of the draw.

TO BELIEVE AND NOT BELIEVE

The more science I learned in graduate school the more probability claims I found. There seemed no question that science accepted the metaphysics of probability, all grounded in sophisticated math "axioms" or assumptions, just as science ignored the mismatch problem and assumed that our gray universe is black and white. I suspected that the black-white and probabilistic assumptions came as a pair. One supported the other, and both lacked external competition. Everything is black or white but we do not know which. So we put odds or likelihoods or probabilities on the black-white alternatives. Only God or No One knows which.

In graduate school I was convinced that we needed a gray or fuzzy science to better model our gray world. The question was whether probability fit in the fuzzy framework. This was the next stage of the hunt. The question was no longer whether fuzziness exists. I had found a math proof that fuzziness exists. The proof showed you cannot equate probability with fuzziness. There are always more measures of fuzziness or types of "fuzzy entropy" than measures of probability. In a later chapter we will look at one such measure of fuzzy entropy and in the last chapter I will use it to look at the big question of why there is something rather than nothing. But for now the problem is where probability fits in. Or does it?

I remember the confused looks on faces when one day at a seminar on neural or brainlike learning theory I sneaked in this question: *Does probability exist?* That was like asking sunbathers on the beach whether the sun exists. Every talk before mine had presumed a probability world. One researcher tried to find the most probable trend line or learning curve that fits a scatter of data points. Another researcher cast learning as the minimization of a probabilistic entropy—brain cells change to increase the information they gain from experience. Another modeled learning as a probability game played by a learning machine or man against a malevolent nature that rewarded or punished actions by increasing or decreasing the probabilities of actions. As their teachers had done before them, these scientists had picked up the probability world view and run with it and published with it and taught with it and passed and failed students with it.

At another meeting I grew tired listening to a physicist make bloated claims about the "scientific method" and how it differed from common sense and from religious paths to knowledge. Physicists love to expound on the "scientific method" as they look down from their press-confirmed vantage point of kings of science. When I took the podium I worked in some comments against probability and in favor of fuzziness. Then out of the blue I asked the big question: Which is easier to believe in, probability or God? No response. No smiles or frowns. I then looked at the physicist, who now sat in the audience. As a professor of electrical engineering I wanted to ask the audience, "What has a physicist done for you lately?" But I knew better and instead just smiled and looked at the physicist and let it fly: "The ultimate fraud is the scientific atheist who believes in probability." About half the audience laughed and the other half grimaced and shook their heads.

In college I had first lost my faith in bivalent science. Now I had gone the next step and lost my faith in the god of probability, the conceptual guardian of the black and the white, the uncertainty hedge that always said "not A but probably A and improbably not-A." Probability looked like fiction. It looked like fiction piled atop the fiction of a black-white universe. No wonder scientists appealed to probability when they criticized fuzziness. If you blew away the fog of "random chance," only an unrealistic world of black-white extremes was left. Probability was the hedge that smoothed out the rough bivalent edges. Probability made the bivalent package reasonable. It sold bivalence. But I did not find it reasonable because I could not find it at all.

I did not dispute the internal math of probability theory. I disputed where the math comes from and how we view it and apply it. The math comes from thin air and we apply it everywhere. Probability math comes from naked assumptions and not from some more general theory. Probability finds its psychological roots in our gambling hunches. So far it has had no logical roots. We can deny the axioms as easily as we accept them. Scientists and mathematicians do not derive the probability axioms from anything. They do not arrive at them by observation or deduction, by experiment or math proof. They arrive at them by will, by stipulation, by arbitrary say-so, by assumption. They like where the axioms lead and so they like the axioms and take them for free. As Bertrand Russell put it, assumption differs from derivation as theft differs from honest toil.

I wanted to know where the roots of probability, the probability axioms, came from. What implied them? Why believe them rather than not? Why the emotional insistence on probability by scientists in every country? What psychology lay beneath the probability commitment? What math lay beneath the psychology? Modern science has few theoretical primitives, few concepts that have no causal or logical antecedents. Gravity remains the central physical primitive. Einstein guessed that energy and matter curve space-time and this curvature we call the force of "gravity." But we still want to ask why and uncover a deeper causal antecedent. Why does matter curve space? Einstein did not prove that matter curves space. He did not derive his curvature equations. He guessed at them. And so far, from the bending of starlight around the sun during eclipses to the strong radiation emissions of massive black holes in the centers of galaxies or buried in ultrabright quasars across the cosmos, it looks as if Einstein guessed well. It appears that matter curves space. But why? We may never know. If we ever find out, the first thing we will do is ask why again and continue the regress. In the meanwhile we experience gravity everywhere and everywhen. We have no trouble pointing to examples of gravity.

But we cannot point so easily to examples of "randomness" or probability. The best we can do is point to the long-term average behavior of flipped coins or lit light bulbs or decaying atomic nuclei. We cannot catch probability in the act. That makes it harder to accept the axioms of probability on faith as unexplained primitives. We not only want to know why the axioms should hold, we are not even sure they do hold. Yet scientists believe them and project the murky psychology of probability onto nature.

As a child I wanted to know what probability was when I first heard of "the odds" and "chance" and the "law of averages." This is where the hunt really started. When I was four years old I played craps with my older brother and his friends. The magic of chance seemed to rule the dice. Yet I knew it was just the hand shaking the dice and the dice bouncing off the lid of a brown cardboard box. At ten or eleven I debated God and the origins of planet Earth with my Catholic cousin. He said God made the world. I said the world evolved "by chance" from small world pieces and gas. In my mind I could see the black void of space filled with round brown clumps of rock and our world just one of the clumps. There might be other clumps where he and I were having or had had the same argument. But I could not make my cousin see it. In my small farm high school in Kansas I often checked out the library's book on probability theory and its gambling history. The book was watered down and the math was too simple to scare or interest me. But the book was my tie to the mystery of being. Then and before and since probability seemed full of magic and wonder and power. It fascinated but confused me. I believed in probability and yet I did not.

My first probability doubts came from the short films we used to watch in science class. I remember sitting in my third-grade class and watching a science documentary. It was about how you grow and harvest and market apples. The teacher ran the film on a 16mm projector. When the reel ended, the teacher rewound the film but forgot to turn something off. So the class got to watch the last part of the film run backward. Apples jumped back up on a tree and water ran uphill. We laughed so hard the teacher let the film run on backward. Somewhere in those images of reverse causality, after the laughter fell off, I imagined running the universe backward, backward to see where I had been, to see if Moses and Jesus had done what the Bible claimed, to see what came out of African trees and ancient oceans and swirling galaxies, to see if Someone or No One had set it all in motion. It was a big brainful. The idea raised a question and has stuck with me ever since: Where did the randomness go? Before the red apples fell off the tree there was a probability that each would fall at a given time in the future. But in reverse the apple-falling probability never showed up.

Probability seemed a psychological side effect of forward-looking creatures, just as free will seems. In retrospect we see many of the causal chains that precede our actions and ideas. Sherlock Holmes would sometimes follow the causal chains a short way into the future and would then startle Mr. Watson by telling him what he was thinking. We gain information as we experience future events. Information up, probability down. Fuzziness works the other way. Information up, fuzz up. More data help us pin down the gray border between thing and not-thing. But probability melts with more data.

That conflicts with the science view that probability lies in the physical nature of things. Most equations in science have no time direction. You can solve the differential equations for past times the same way you solve them for future times. Time looks like an illusion too. In that sense you can run the universe in reverse as easily as it runs forward "by itself." That takes the hunt to the next stage. If probability does not lie outside our skulls in the nature of things and yet we think it does, where else can it lie?

THE PROBABILITY INSTINCT

It looks as if Kant was right: He thought our minds structure our perceptions. Probability was built into our minds. Our minds, the electrochemical symphony that our narrowly evolved neural ganglia play, impose an infrastructure on our thinking. The mind imposes a background of time and space and causal connectedness. Scientists have never seen a "causality" in the wild. They have seen, and they predict, only space-time events that follow space-time events. Apples on the tree, then apples in the air, then apples on the ground. Equations and correlations have replaced causes, just as science has largely replaced philosophy and religion as a theory of things. No causal germ in one event unfolds into another event. But the mind, as eighteenth-century philosopher David Hume observed, makes it seem so and inserts the causal links in the event chain.

Probability seems to be part of the same mental infrastructure. It forms part of our mental background or viewing screen along with time and space and causality and similarity and the topological notions of continuity and connectedness. We see probability everywhere because it lies in our glasses.

I believe that probability or "randomness" is a psychic instinct or Jungian archetype or mental trend that helps us organize our perceptions and memories and most of all our expectations. Probability gives structure to our competing causal predictions about how the future will unfold in the next instant or day or season or millennium.

Probability ranks or weights the future alternatives. Our expectations then blend or average these future alternatives into a single probability-weighted average. The probability weights do not exist outside our minds. They have no physical reality but have a powerful psychological reality rooted in our neural microstructure. Hume also thought that we make up probability as we go and use it to fill in gaps in our mind schemes or world views: "Though there be no such thing as *chance* in the world, our ignorance of the real cause of any event has the same influence on the understanding and begets a like species of belief."

This *probability instinct* seems to cut across cultures and may cut across species. Besides the probability-laden psychology of scientists and most nonscientists, the widespread gambling and games of chance in primitive and modern cultures suggest that probability "reasoning" may be a cultural constant like hero worship or fertility rituals or incest and adultery taboos. A cultural constant suggests a biological substrate, and that requires an evolutionary history.

Ranking future alternatives can help pass on genes. Those who could so rank may have eaten those who could not. It allows us to bet before we act and improve the outcome of acting. That forward-looking ability has supreme survival value in biological evolution, the genetic variation and selection in the last few million years that has finely sculpted our brains and minds, and in the prior evolution that sculpted the brains and minds of our mammalian ancestors in the last 220 million years. Natural selection filters out organisms as they cross the fuzzy line from the present to the future. Natural selection favors brain mechanisms that help an organism make its next move in a changing and dangerous world. These forward-looking brain mechanisms may run deep in the structure of mammalian and even reptilian brains. Future studies may find that the brains of chimps and apes and lesser-brained mammals house a forward-looking probability instinct. At the other extreme we should not be surprised that scientists have exalted probability ranking into their grand organizing principle of maximum probability. Scientists follow their probability instincts as their hominid forefathers followed theirs. Scientists just know more math.

SUBSETHOOD:
To Contain and not Contain

A probability instinct explained the psychology of scientists but did not explain where the probability axioms came from. It gave no insight into the formal nature of probability. To know that you had to know what parent begot the probability child. Whose baby was it? So the last stage of the hunt was a search for parents.

I found a parent. Other parents may well exist in the infinite labyrinths of math. But I found at least one parent of probability and it is pure fuzz. It is so fuzzy that earlier fuzzy theorists missed it, as had I. We missed it because we presumed it was bivalent and so we never looked carefully at it.

We had overlooked the fuzzy idea of *containment,* how much one thing contains another thing, how much one set contains another set. Fuzzy sets arise when a set partially contains an element, as when an audience contains a somewhat happily employed person or when a barrel contains a somewhat rotten apple or when a chromosome contains a somewhat mutated gene. In some sense the set is not fuzzy but its elements are fuzzy. The elements all have some property to some degree. I call this *elementhood*. The old fuzzy or multivalued logic is all about elementhood.

What happens when one set contains another set? Put a little box in a big box. Then the big box contains the little box. It contains the little box 100%. Can that containment take on degrees? Sure it can. Put the little box only halfway in the big box. Then the big box contains the little box only 50%. The little box is both in and outside the big box.

I called this fuzzy containment *subsethood*, the degree to which one set is a subset of another set. Traditional fuzzy theory assumed subsethood was bivalent, all or none, 100% or 0%. That seemed as extreme as any other black-white claim. Very tall men made up a 100% subset of tall men. That I could buy. Every very tall man is tall. But the old view said that tall men made up only a 0% subset of very tall men. That I could not buy. It was a matter of degree. Every tall man is very tall to some degree, often to a very slight degree.

In hindsight subsethood was the next step from the elementhood of fuzzy sets. Elementhood puts balls in boxes. It puts them in to some degree. Subsethood puts boxes in boxes. In the special case the input box shrinks to a ball and you get back the old idea of a vague or fuzzy set whose elements belong to it to some degree. So subsethood subsumes elementhood as a special case. Yet subsethood differs from elementhood. Subsethood holds *between* sets. Elementhood holds *within* sets.

The between-set nature of subsethood implies that subsethood holds for regular bivalent sets too. Aristotle can govern the bivalent structure within the sets but the Buddha, so to speak, still governs the structure between them. We can grab any two sets in the world of math and ask to what degree one contains the other. In the bivalent framework we cannot ask that. In most cases neither set totally contains the other. That means no one had looked down this corridor. Bivalence had shut the mental door. I opened the door and that is where my hunt for probability became a math hunt.

THE ZEN OF GAMBLING:
THE WHOLE IN THE PART

A new theory of probability must account for our gambling experiences. It must give back the old formulas that we use to calculate odds and to place bets. It must give them back as special cases of some more general formulas. Could subsethood do that? I didn't think so. Partial containment seemed a purely fuzzy matter outside the black-white world of probability. I was wrong.

Most of us learn about probability by flipping a coin to settle a dispute or to pick who goes first. Or we play bingo or poker or the horses. Our probability instinct resonates with the behavior of Las Vegas gaming tables and insurance-policy pricing and bouts of Russian roulette. The axioms of the Russian probabilist Andrei Kolmogorov do not compare with the experience of rolling the dice with some blue chips on the table.

Relative frequency lies at the heart of gambling and our probability instinct. Flip the coin 100 times and it comes up heads 46 times. Then the relative frequency of heads equals 46 per 100 or 46%. In general *relative frequency equals successes divided by trials*. Every trial is a success or a failure, a win or a loss. So the number of trials equals the number of successes plus the number of failures. For this reason statisticians sometimes refer to relative frequency as the *success ratio*, the ratio of heads to flips, baskets to shots.

Where lies fuzziness in the gamble? This question threw me for years. By design all events in the probability framework are black or white. Aristotle's excluded middle law of A OR not-A always holds. The flipped coin comes up heads or tails and not on its edge. The outcome is bivalent, 1 or 0. Sets of trials or flipped coins are bivalent too. If there are 46 successes out of 100 trials, then there are 54 nonsuccesses or failures. The set of successful trials does not overlap the set of failure trials. And the two sets exhaust or fill up the set or "space" of trials (Figure 3.2).

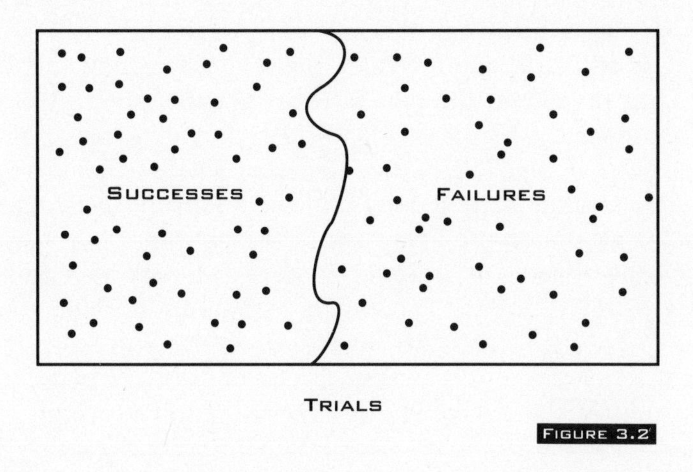

FIGURE 3.2

The black dots stand for trials like coin flips or target shots or basket shots.

Where is the fuzziness in this Venn diagram? A OR not-A holds so where can it be? The answer lies in how much one set contains another, in subsethood. But where? The set of successes has nothing in common with the set of failures. The two are disjoint. We can draw a hard line between them. Neither set contains the other to anything but a zero degree. The only other set around is the set of all trials, the entire rectangle. This set too is bivalent. It contains all its subsets to 100% degree. The whole always and totally contains its parts. In particular the set of all trials totally contains the set of successes and the set of failures. The two pieces of the rectangle are parts of the whole rectangle. So where is the fuzzy containment?

I came back to this question over and over. What good does subsethood do here? All the sets of interest contain one another all or none. The parts do not contain one another at all. The whole totally contains the parts. What's left? There was only one possibility left and it made no sense. It had to be wrong. I kept ignoring it as an error. The very idea sounded like bad Eastern mysticism. My probability hunt seemed to have ended in nonsense. But the idea would not go away. It was the only alternative left. I circled back to it many times. At last I decided that maybe my intuition was wrong and the idea was right.

The whole in the part. Every whole contains its parts. But do the parts contain the whole? On the face of it no. The question sounds absurd. How can the part contain the whole? The one exception is the degenerate case when the part equals the whole itself. But in general the part differs from the whole. Here the part cannot totally contain the whole. But it always *partially* contains the whole. The part contains the whole to some degree. Consider another Venn diagram, of the nonfuzzy set A that sits in the big set or "space" or whole X (Figure 3.3).

The shaded set A is a subset of X. It is a part of X and it is not fuzzy. It has nothing in common with its opposite or comple-

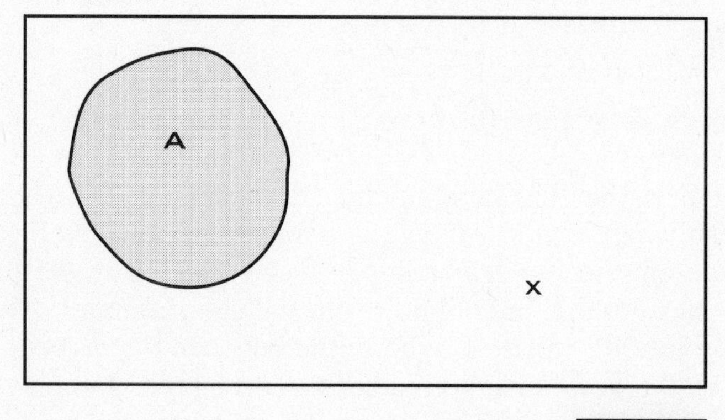

FIGURE 3.3

ment not-A, the unshaded rest of the rectangle. Here A stands for the set of successes in Figure 3.2.

Now consider a thought experiment. Suppose the shaded set A shrinks to a point and vanishes into nothingness or the empty set. In that extreme case the part, the empty set, clearly does not contain the whole at all. Nothing cannot contain something let alone everything. In this case the containment or subsethood is 0%. Now suppose A comes back into existence and grows. Eventually A grows as far as it can and fills up the whole rectangle. Then the part A equals the whole X. In that extreme case the part contains the whole 100% because the part is the whole. What happens between the extremes? Containment takes on degrees. It varies smoothly from 0 to 1 as the region A grows in area. The bigger the subset A, the more it contains the whole.

So there it was. The part cannot totally contain the whole unless the part equals the whole. That much, and only that much, scientists had gotten right. Otherwise the part partially contains the whole.

The part contains the whole in direct proportion to its size or mass or overlap with the whole. That they had not gotten right. For centuries scientists and mathematicians had missed this simple idea. They missed it for the same reason I had missed it: Aristotle had outlawed the very idea of it. We just assumed containment was all or none, white or black, bivalent. We never checked that assumption. We filtered the gray right out of our glasses. And gray holds in almost every case.

Now comes the punch line. What is the whole in the part? The concept is new but general and must have shown up in many places in formal reasoning and math and in our daily experiences. When it showed up we did not have an explanation for it. We had to take it as a theoretical primitive and justify it by how it worked and the things it led to, by its utility in practice and its fertility in theory. We gave it names and built near-mystical intuitions around it and we saw instances of it wherever we looked.

The whole in the part is probability. It is the probability of the part. What is the probability of success? The degree to which all trials are successful, the degree to which the set of successful trials contains the set of all trials. A quick math proof shows that this odd subsethood relation equals the relative frequency that we all use but do not derive.* In general the probability of a set or event A equals the degree to which the part A contains the "sample space" X. The probability of A is how much the whole X sticks in or fits in the part A.

All this gets wrapped up in what I called the Subsethood Theorem (which shows that subsethood degree equals what probabilists call a "conditional probability" like the probability that the tree falls if you cut halfway through its trunk). This theorem rests on a direct extension of the old Pythagorean Theorem for right triangles. The Pythagorean Theorem, which gives the length of a rectangle's diagonal as the square root of the sum of squared lengths of two sides, is the most important theorem in math. If in 500 B.C. or so old Pythagoras had looked in the right direction, he would no doubt have seen this extension and might have launched fuzzy theory more than a century before Aristotle was born. The terms "probable" and "random" might never have survived the next two or three millennia of cultural evolution. Today we can call the whole in the part the "probability" of the part but that label adds no cognitive content. The quantity in question is a degree of subsethood. It has been so since before the Big Bang and it will remain so after the Big Crunch, at least if math still holds after a universe collapses into nothingness. "Randomness" is dead because it never lived. The hunt ends in a kill.

The proof of the Subsethood Theorem came to me while I soaked in a hot tub. That's when I saw its geometry. The math details were simple after that insight. The Subsethood Theorem was pivotal in my geometric work on fuzzy theory and pivotal in the direction my career and world view took. For that reason I want to relay the genesis of the Subsethood Theorem.

In late October 1989 I was rewriting my paper "Fuzziness *vs.* Probability" for the third time. I came to the subsethood section and stopped cold. I could not get the geometry right. I could not get the geometry at all. I sat and looked at the computer screen in my home office and thought about taking a nap on the

---

*Let S(X, A) stand for the degree to which X is a subset of A. Since A is a total subset of X, S(A, X) = 1. But in general the converse subsethood S(X, A) falls between the bivalent extremes: $0 < S(X, A) < 1$. Suppose X contains N trials and A contains the N_A successful trials. Then the Subsethood Theorem implies that

$$S(X, A) = \frac{\text{(number of elements in } A \cap X)}{\text{(number of elements in } X)} = N_A/N,$$

the relative frequency of successes in trials.

waterbed wedged in the other corner of the room. But I did not nap. It was still early in the evening and a full workout lay ahead. I wrestled some more with the geometry and sketched unit squares on the back of laser-printed copy. I could feel the answer in me as you feel a tip-of-the-tongue name come to you after a question or as you feel a comeback line come to you after an insult. A solution has a way of bubbling up out of your subconscious if you brood about a problem long enough.

I worked out at home and went to the gym to ride the LifeCycle. I rode it for 50 minutes and closed my eyes and meditated through ten minutes or so of the sweatier part near the end of the ride when the beta-endorphins really flow. Exercise perturbs concentrated thought, pounds it out of your head before it takes seed. Sometimes in the middle of a hard bike ride I try to state and prove simple math theorems, like the quadratic formula. But even that is like filing papers outdoors in a windstorm.

I came home from the gym and went straight to the hot tub behind my house in the avocado grove. The aerobic stress had oxygenated my brain cells and pumped up the endorphins to maximum levels. A large number of creative ideas have come to me as I soaked in a hot tub after a hard workout. Almost none have come to me in a formal office. Creativity and responsibility don't mix.

A hot-tub bath after a hard workout brings with it some of the workout high. That is when my head is clearest and when I can sit the longest without pain in the full lotus position with legs crossed and the feet high up near the hips.

I sat in a quiet *zazen* meditation. In *zazen* you do not chant or count your breaths or do any of those types of meditation with training wheels. No guru sells you a mantra to repeat over and over and you do not test New Age hypotheses. You simply sit still and think still. You try to prove B. F. Skinner wrong, that you can do more than respond to the environment outside your skin and the environment within your skin, that you can do nothing and "blank your mind" and look for the empty set.

You just sit still and let the thoughts pass by and dissolve like smoke puffs. *Zazen* takes more effort than hard exercise and it cleanses better if you do it right and don't cheat. *Zazen* costs nothing and no one can ever take it from you. When gravity has won and you have lost your looks, muscles, figure, posture, hair, teeth, energy, sex, health, income, friends, and family—when the external stimuli have gone and the internal stimuli hurt—you will still have *zazen* if you want it. But then you may not be able to sit in full lotus position.

I was having an empty *zazen*, the best kind. I don't know how long I sat there sweating and staring at the chaotic froth and swirling vortices of steam on top of the water. But then the intruder came and ruined my *zazen*. My mind's eye was supposed to be blind now but it was not. I "saw" a gray waterbed hanging in midair above me as if in a surrealistic painting. How *close* is it, I asked myself or my nonself. At this point the *zazen* collapsed and I was in pure math concentration. In the next instant I held several white strings in my right hand and all tied to different parts of the waterbed's surface. It resembled flying a kite or landing in a parachute. How close is the waterbed? The shortest string, of course. Which string is that? The *perpendicular* string. The right triangle. The string that defines a right triangle with the string as one leg, with another string from me to the corner of the bed as the hypotenuse, and with the line segment from that corner to where the first string hits the waterbed as the other leg. An "orthogonal projection" as they say in math. That was the answer. No question. Perpendicularity. Orthogonality. The Pythagorean Theorem.

I did not jump up and shout "Eureka!" like Archimedes did when he saw how much water he had displaced in his bathtub. I was struck with the fear that can lead to panic, as when you realize you left your wallet on the checkout counter. I had published an algebraic solution to the Subsethood Theorem and had repeated that solution in the first part of my "Fuzziness *vs.* Probability" paper. But that subsethood measure differed from this one. And this one was right. The thought of publicly repudiating my own work on subsethood, with a chunk of it in my Ph.D. dissertation, began to outweigh the thrill of discovery. I jumped up out of the hot tub and ran to the house.

I went back to my downstairs office and lay on the waterbed that had earlier seemed to float in midair. I wrote down the new geometric measure of subsethood and compared it with the old algebraic measure. I put the same values in each measure and got the same answers out. That could mean they were the same. I remember my heart pounding in the excitement. More test values led to more same answers. But then tragedy struck: another test value led to two different answers. It was all over. My heart now pounded for different reasons.

I had computed the answers by hand and so I redid them to be sure. This time all the answers came out the same. The apparent difference was my mistake. Despair turned to triumph.

I guessed that the two subsethood measures were in math fact the same thing. In a few more minutes I had proved it as a theorem. The new equaled the old. The world was just. The two subsethood measures were the same thing but in different notation like the same message in English and in Chinese. The geometric measure was simpler and rested on fewer assumptions. I carefully wrote out the proof again, this time in black ink and with lots of space between the symbols and the lines. The two subsethood measures were still equal. I had proven, or in some sense reproven, the Subsethood Theorem and from fewer assumptions. I had gotten the geometry right and the theory was complete.

A little later I worked out some corollaries of the Subsethood Theorem. One corollary shows that subsethood reduced to probability in the special case of the whole in the part. That was quite a bonus. I called and faxed my math friend Fred Watkins, who runs the fuzzy software company HyperLogic in Escondido. I faxed Fred the new theorem and proof and remember telling Fred the fuzzy joke: What does the fuzzy theorist say to the probabilist? Thanks!

The whole in the part. That is exactly as it should be. The very idea connotes the antiscientific beliefs of Eastern mysticism that scientists deplore. But it is science. It is pure and elementary math and the Buddha wins that round. And Einstein looks right again: God need not play dice. The universe is not random. You can take it one step deeper and get rid of the "randomness." The universe is deterministic but gray. Chaos theory had already gotten the determination part right. Fuzzy theory now confirmed that and should that all things were matters of degree too.

My quest was over. Once again scientists had erred at the "self-evident" level of logic and math. They had blown up their hunches and instincts and conditioned reflexes of "probability" and "randomness" into a pagan god and packed them into every corner of the universe just as a century before they had everywhere packed the invisible "luminiferous ether" so that light waves could move through space. I had lost faith in science. I could not trust scientists again. I could not even trust my own instincts and intuitions. The whole in the part showed me that.

So who can you trust? A fuzzy theorist can give only one answer: Don't trust me.

II

The Fuzzy Past

4

THE FUZZY PAST

There is nothing new under the sun. Is there anything of which one can say, "Look, this is new"? No, it has already existed long ago before our time.

ECCLESIASTES 1:9–10

All ideas come from other ideas. Fuzzy logic did not burst ready made onto the world of science and engineering, though the popular press has written it that way. The name "fuzzy logic" is cute and recent and we are stuck with it. "Fuzzy logic" is less formal and less exact than the old term *vagueness*. And it does not clot in the mouth as does the formal term *multivalence*. The name labels an idea or family of ideas—shades of gray, blurred boundary, gray area, balanced opposites, both true and false, contradiction, reasonable not logical—and those ideas are very old and have many ancestors.

This section looks at where fuzziness has cropped up in ancient and recent history and who has refined or damned it. I have focused on the early twentieth century because then much of modern science, math, and philosophy crystallized or ossified. Russell's new logic and Heisenberg's new quantum mechanics shook science and forced scientists to rethink math and science and how they described the gray world with black-white tools and how much it cost them. The ice had cracked and fuzziness seeped in. A modern generation had to face fuzziness for the first time since ancient Greece and had to accept it or deny it. To accept it meant they had to overturn a world view. To deny it meant constant footwork and word play to keep the old holes plugged while new holes sprouted.

The ancient history of fuzziness reduces to the logic of the West and the East. In the West Aristotle gave us our binary logic and much of our scientific world view. He taught us to logic chop and always draw the line between opposites, between the thing and the not-thing, between A and not-A. The better you drew those lines, the more logical your mind and the more exact your science.

In contrast the great cultural leaders of the East were "mystics." They tolerated ambiguity or vagueness and even promoted it. The Buddha rejected the black-and-white world of words on his path to spiritual or psychic enlightenment, while Lao-tze gave us the Tao and its yin-yang emblem of opposites, both thing and not-thing, both A and not-A. History might have been different, and our world today might be very different, if the Buddha and Lao-tze had learned the math and logic of the ancient Greeks.

5

ARISTOTLE VS. THE BUDDHA

Everything must either be or not be, whether in the present or in the future.

ARISTOTLE
DE INTERPRETATIONE

I have not explained that the world is eternal or not eternal. I have not explained that the world is finite or infinite.

THE BUDDHA
MAJJHIMA-NIKAYA

The fundamental idea of Buddhism is to pass beyond the world of opposites, a world built up by intellectual distinctions and emotional defilements.

D. T. SUZUKI
THE ESSENCE OF BUDDHISM

Aristotle and the Buddha meet in fuzzy logic. The prophet of A OR not-A confronts the prophet of A AND not-A at the level of math, science, and engineering. This may be the first clash of Eastern and Western belief systems at the technical level. It will not be the last.

The United States and Japan lead the world in business and technology—in money and math. Both countries have cultures derived from other countries. Ancient Greece stands to the

United States and much of Europe as ancient China and India stand to Japan.

Aristotle and the Buddha personify these two cultural roots. The Buddha was Indian and not Chinese and never went to China. But his world view passed through China. It passed through the filter of Chinese Taoism and ended in the Zen Buddhism that permeates Japanese thought and culture and history and business practice. This chapter looks at some of the ancient and recent influences of these two great thinkers.

THE SOCIOLOGY OF CREEPING FUZZINESS

How much do we in the West know about Aristotle and the Buddha? How much have these men changed world culture?

Aristotle's logic and scientific bent have shaped much of the modern Western mind and defined its range of parameters, its boundary, its list of the correct and incorrect. Each generation has refined Aristotle's model of the mind and the universe. Science has given up Aristotle's view of the universe. That started with Sir Isaac Newton's theory of white light as a mix of colored light. Aristotle also used a prism to separate white light into colors. But he said metaphysical "transformations" caused the separation. Unlike Newton he did not point to a physical mechanism like a wavelength filter.

Science has been less critical of Aristotle's model of the mind. To a large degree Aristotle still defines what is *philosophically correct* in logic and reasoning. How much confidence would we put in our Aristotelian boundary if we thought Aristotle was wrong or he wrote on whim or he only reaffirmed the fleeting pop culture of his ancient day?—if accepting black-white logic had more to do with taste or whim or cultural conditioning than a "self-evident" logical "necessity"? If we don't accept Aristotle's philosophy in our physical theories, why should we accept it in our reasoning and our computer design? What lies outside our Aristotelian boundary? What about multivalued or "fuzzy" logic? How do we explain the creeping fuzziness in the high tech of Eastern countries? Where does fuzzy logic fit in? Where does it fit in Eastern belief systems?

In the early 1990s fuzziness has emerged as the technical and cultural emblem of the Far East. Japan has led the fuzzy revolution at the level of high-tech consumer products. Japanese engineers have used fuzzy logic to raise the machine IQ of camcorders and transmissions and vacuum sweepers and hundreds of other devices and systems. The Japanese government has set up two large labs and each sponsors fuzzy conferences in alternate years. I have watched Japanese TV commercials for smart washing machines and air conditioners in which the only word I understood was *fuaji* said with an exclamation mark. The Japanese have invented new Kanji characters for "fuzzy theory" (*fuaji riron*). Subway riders and corporate executives read the many popular Japanese books on fuzzy logic. Japanese TV has run prime-time specials on fuzzy engineering and how it applies to manufacturing and consumer electronics. Politicians in the Diet have debated and joked over the meaning of fuzzy logic. MITI, the Ministry for International Trade and Industry, estimated that fuzzy products, 70% of them in consumer electronics, accounted for roughly $1.5 billion in revenue in 1990 and over $2 billion in 1991. The global market for computer services, software, and hardware is about $200 billion per year. That puts the Japanese fuzzy effort at about one percent of the global computer market. And the fuzzy race has just begun.

Fuzziness has taken root in other countries in the Far East. Each country expresses its fuzzy world view in its own way. Japan has focused on practical engineering. South Korea has followed suit with its own fuzzy society and conferences and corporate efforts as it competes with Japan. Singapore, Malaysia, and other Southeast Asian countries pursue fuzzy logic in their universities and in new start-up firms. India has produced several fine fuzzy theorists. Taiwan and Hong Kong have also spawned start-up firms but they seem to have balanced much of the engineering side of fuzzy logic with its more philosophical and mathematical expression in China.

When I visited Beijing in the spring of 1989, just before the Tiananmen Square massacre, I learned that China has over 10,000 fuzzy theorists and students and has several technical journals on fuzzy math and engineering. Chinese engineers have applied fuzzy systems to manufacturing and military tasks that include the control of the optimum thickness of sheet plastic for wrapping industrial goods and the smoothing out of navigation and guidance control in fighter cockpits. My host, Professor

Wang Peizhang, at Beijing Normal University, has founded from a distance the company Aptronix, the first fuzzy start-up company in the Silicon Valley. Professor Wang grew up in the pre-1949 culture of classical China and mastered advanced math and the Chinese art of meditative breathing and the even finer art of academic politics. In the late 1960s, during Mao's Cultural Revolution, Wang weeded turnips in the fields. Two decades later he became chairman of the board of Aptronix. Wei Xu, the president of Aptronix and former Chinese government bureaucrat, turned 30 in 1992, the year *Business Week* featured him as the new Steve Jobs of fuzzy logic.

To understand the triumph of fuzziness in the Far East you must understand the "Oriental mind," as my Far Eastern friends tell me. Much of that depends on understanding Buddhism, its history and practice and how those two differ.

I believe Buddhism correlates with fuzzy engineering in countries with developing high-tech economies. I can state this sociological hypothesis in color. Consider two globes of the modern world. Take the first globe and color red the countries that pursue fuzzy logic in their science, engineering, and math. Take the second globe and color red the countries where Buddhism flourishes. Then you will have two similarly colored globes.

ARISTOTLE AND THE BUDDHA IN THE WEST

The Buddha died in India nearly one hundred years before Aristotle was born in Greece in 383 B.C. Both men reflected, codified, and extended the ideas of their day. The Buddha embodied the wisdom of the East. Aristotle embodied the wisdom and mathematic-scientific outlook of the West.

Alexander the Great, Aristotle's pupil, conquered deep into India, which today bears the Buddha's wheel of truth (*dharma*) on its flag. Alexander no doubt saw the emerging Buddhist order amid the ruling Brahmin order of India. Aristotle may have learned something of Buddhism before he died in Greece in 322 B.C., a year after Alexander died in Babylon. As an aside, in 1894 the Russian explorer Nikolas Notovich published a book that claimed that in Tibet he had found evidence that the child Jesus walked to India or Tibet along the Silk Road during his "lost years" from age 12 to 30 and there studied Buddhism.

The Buddha and Aristotle set in motion causal chains that wind through Eastern and Western history and still have not really met and intertwined. You find Buddha sutras in Japanese hotels and Bibles in American hotels.

Ask an Easterner about Christianity or Judaism or Islam and you will get an informed answer. Ask a Westerner about Buddhism and far more often than not you will get a Hollywood caricature, a fat man sitting in full lotus position with sitar music playing in the background and one hand clapping and a mouthful of love-love-love slogans and the idea of nirvana cast as an infinite orgasm. Imagine the Christian response to a like depiction of Jesus in movies or books or restaurants, or to a Japanese or Cambodian or Indian version of the TV series *Kung Fu* with a roving Christian monk kicking and punching the hell out of the locals.

How many Christians or Jews or Euro-American atheists or agnostics believe that an Asian politician or journalist can grasp world affairs, let alone world history, without some understanding of the Bible? As a child or adult you may have read the Bible. You certainly saw the movies. You saw Moses part the Red Sea, David bring down Goliath, Jesus crucified and resurrected. You may have read parts of the Koran or the Gemara or looked at the blue-colored pictures in somebody's book of the dead or glanced at a Bhagavad Gita handed to you at the airport.

But imagine the poll results: How many Westerners can state the basic tenets of Buddhism? Can list three out of four Noble Truths? Can name and describe a single sutra? Can tell the Buddha from Confucius and from Lao-tze? Can place them in history? *How many Westerners can name five Eastern books?* At which Academy Awards ceremony did the epic spectacle of the Buddha's life and times sweep the Oscars? How many movies have you seen of the Buddha, or of Asoka's unification of India and his battlefield conversion to Buddhism, or of Buddhism's clashes with Hinduism or Islam or Communism or its spread to China, Korea, Southeast Asia, and Japan?

I wonder to what degree our Western view of Buddhism would have changed if composer Richard Wagner had lived one or two more years. Wagner was the writer-director-loudmouth artist superman of his day. In 1883 he died at work on another grand-opera music drama, *The Victors*, the tale of how the Buddha's disciple Ananda gives up a woman's love to sit and meditate at his master's side.

Aristotle and the Buddha lived very different lives. Aristotle was a man of court as well as of learning. Plato trained Aristotle for 20 years in the Academy. Later Aristotle taught princes. Aristotle reduced the workings of the universe and society to metaphysical pieces. (The word "metaphysics" arose because early chroniclers and lecturers followed Aristotle's works on physics with his works on ontology, hence "meta"-physics or "after"-physics.) In his multivolume *Summa Theologica* and *Summa . . . Contra Gentiles*, thirteenth-century philosopher and theologian St. Thomas Aquinas polished, debated, and extended the works of Aristotle. Aquinas refers to Aristotle as "The Philosopher" throughout those many volumes. In some sections almost every page cites him. The Catholic Church owes a great deal to Aristotle. The fourteenth-century Italian poet Dante called him "the master of those who know."

Aristotle was politically correct by the standards of his day but not by ours. He argued for the legitimacy of aristocracy and slavery and may have kept slaves. In *The Politics* Aristotle tells us that "there exists a legal or conventional slavery. This arises under the convention that provides that all that is captured in war becomes the legal property of the captors." So if you or someone else conquers and enslaves a person, you can legally sell or buy the person—a legal principle that stayed active through the American slave trade. Aristotle later claimed that "there is a difference between the rule of master over slave and political rule. Rule over free men is by nature different from rule over slaves," and proceeds to make one of the most unabashed cases for might makes right in political history.

Most of all Aristotle wrote things down. Most volumes have been lost but enough have survived to shape Western history. Aristotle wrote several volumes on logic and reasoning in *The Organon*. He wrote on all the branches of philosophy and science and art of his day. His *Poetics* argues for the three-act structure that underlies every Hollywood two-hour screenplay (Act 1 = 30 minutes, Act 2 = 60 minutes, Act 3 = 30 minutes) and hour or half-hour teleplay. Some modern screenwriting courses assign the *Poetics* as required reading.

Aristotle seems to have lectured and worked in Greek social and military life less than did Socrates and Plato. Aristotle wrote more than he preached.

THE BUDDHA IN THE EAST AND TODAY

The Buddha only preached. The legend says that at age 28 he walked away from the wealth and power of his princedom—his father was a local king or *raja* in northeastern India—and never walked back. Most of us walk in the other direction. That has always impressed me more than has anything else about any major religious figure. No one feels the vow of poverty like a young playboy prince who has his health and a daily and free sensory smorgasbord. We have no comparable example in the West. Andrew Carnegie donated a large part of his wealth in his old age but seems to have done so out of remorse or repentance. Today we have many rich men stripped of their millions but someone else always does the stripping.

The Buddha wandered as a beggar to spread his message of renunciation. He did not let followers write down his words, draw or engrave his image, or exaggerate, romanticize, or deify him. Apparently scholars recorded the numerous Buddha canons in the second or third century A.D. A lot changed in those 600 or so years of oral record. A personal philosophy became the religion Buddhism, and a man become more than a man—in modern terms the Buddha was most likely an atheist or agnostic. A lot has changed since then too.

Buddhism spread east to Asia from India and split into little-raft and big-raft species, the schools of Theravada and Mahayana Buddhism. The every-man's-on-a-raft school of Theravada Buddhism still dominates in Sri Lanka and Southeast Asia. The we're-all-on-the-raft-together school of Mahayana Buddhism spread through Tibet, China, Korea, and Japan. Theravada Buddhism focuses more on the self-oriented things the Buddha said. Mahayana Buddhism focuses more on the others-oriented things the Buddha did. Both schools have deified the Buddha and wrapped his teachings in the very ornate rituals and practices that the Buddha struggled to prevent. The two rafts have drifted far.

I took a ride on the big raft to see if I could recognize Buddhism among the Buddhists. By "chance" I moved into a house in the Puente Hills in California less than a mile from the largest Buddhist temple in the Western Hemisphere. In November 1988 the Taiwan-based Hsi Lai Temple opened for spiritual business. Soon after the Dalai Lama came by for a visit. Overnight prices of houses shot up, mine included. Local protests turned to pride in cultural diversity and calls for press and tourist recognition. Real estate agents were quick to define the boundary of the temple's "holy radius" and to include "temple view" in property listings.

I toured the temple shortly after it opened. It had more gloss and opulence than any Buddhist temple I had seen in the Far East. On a clear day the Hsi Lai Temple looks out from the Puente Hills over the San Gabriel Valley to the tall and sharp San Gabriel Mountains. Inside the temple large golden laughing buddhas and bodhisattvas sit surrounded by thousands of smaller buddhas sitting in small golden eggs. Beneath the big buddhas and bodhisattvas the Diamond Sutra lecture sprawls etched across a black stone wall. A museum houses donated buddha statues from around the Orient. Bald nuns in gray and brown robes run the temple and gift shop. Ornate buildings that suggest Beijing's Forbidden City enclose a large courtyard of offset cement rectangles with grass growing between them. The rectangles stand for the thousand steps from humanhood to buddhahood and lead up to the temple's central hall, a vast worship area with more buddhas and bodhisattvas and buddha eggs and a huge golden chandelier that I once stood under and wondered about as a small earthquake shook the hall. Early in the morning worshippers in long black robes line up in the courtyard, chant from their liturgy books, and walk with their hands cupped and heads bowed into the central hall for more chanting and for religious instruction. It is all pleasant to look at but far removed from a man who would not let his followers write down his words or draw or sculpt his image.

The Hsi Lai Temple succeeds in a way its founders did not intend. Each year it draws more visitors, but as tourists and curiosity seekers, not as Buddhist converts. The appeal of

Eastern religions has always been their *experiential* component, not the ritual and make-believe wrapped around it. Western religions offer plenty of that. The Maharishi Mahesh Yogi pulled off something of a leveraged buyout of Hinduism when he removed mantram yoga from its religious shell and called it "transcendental meditation" and sold meditators their personal secret mantras to chant silently over and over for 20 minutes a day. In much the same way karate, judo, and the other martial arts have flourished in the West when stripped of their historical and cultural contexts. Buddhism has so far resisted a Western popularization, perhaps because what a Westerner sees of it is only ritual and statues and the *Kung Fu* television series.

So what does Buddhism say? What were the secular teachings of the Buddha? The Buddha reduced his world view to four points: (1) life is suffering (*dukha*), (2) suffering arises from desire (*tanha*), (3) eliminate desire and you eliminate the suffering, and (4) live a decent life and meditate to help eliminate desire. Want not, hurt not. This is less religion and more an intellectual pain pill.

The Buddha hammered these points over and over in his recorded conversation or sutras. He refused to get caught up in words, the "world built up by intellectual distinctions and emotional defilements" as Buddhist scholar D. T. Suzuki described it.

The Buddha was not a fuzzy theorist in a mathematical sense. He wrote no papers on fuzzy sets or systems. But he had the shades-of-gray idea: He tolerated A AND not-A. He carefully avoided the artificial bivalence that arises from the negation term "not" in natural languages. Hence his famous line: "The no-mind not-thinks no-thoughts about no-things." The Buddha seems the first major thinker to reject the black-white world of bivalence altogether. That alone took great insight and detachment and tenaciousness in an age with no formal analysis. He built a personal philosophy atop his rejection of bivalence. Today we in the West associate Buddhism with the big-bellied caricatures of that personal philosophy.

The Buddha refused to let words get in the way of what matters as a living and dying organism. Avoiding black-white boundaries helped one see the connected world more clearly and focus on the lot of man. The quote from Buddha at the start of this chapter continues, ". . . I have not explained that the world is finite or infinite. And why have I not explained this? Because this profits not, nor has to do with the fundamentals of religion, nor tends to aversion, absence of passion, cessation, quiescence.''

The Buddha focused on death and the old age and suffering that tend to precede it. There was more of that in his day than in ours and we have other painkillers. But life still ends quickly and badly. The Buddha wins at the boundary.

6

What Is Truth?

Let us grant, then, that the deviant can coherently challenge our classical true-false dichotomy. But why should he want to? Reasons over the years have ranged from bad to worse. The worst one is that things are not just black and white; there are gradations. It is hard to believe that this would be seen as counting against classical negation; but irresponsible literature to this effect can be cited.

Willard Van Orman Quine
Philosophy of Logic

Nothing in the world, no object or event, would be true or false if there were not thinking creatures.

Donald Davidson
Journal of Philosophy, vol. LXXXVII, no. 6, June 1990

Do not confuse the moon with the finger that points at it.

Zen proverb

All you have to do is write one true sentence. Write the truest sentence that you know.

Ernest Hemingway
A Moveable Feast

What is truth? This question seems the supreme question of philosophy and science and the pursuit of knowledge. Science seeks truth. Art seeks beauty. Society seeks justice. Truth tops the list of these three noble pursuits.

Truth is also where formal fuzzy logic begins. The *mismatch problem*—gray world but black-white scientific description—reduces to a truth problem, the problem of gray truth. The bivalence of modern science ignores or denies or whitewashes and blackwashes gray truth. That tactic leads to paradoxes and self-contradictions. The fuzzy view says almost all truth is gray truth, partial truth, fractional truth, fuzzy truth. It lets math truths remain black or white as extreme cases of gray.

Fuzzy logic says *all* scientific truths are gray. Bivalent science says none are gray. It says the truths of science are not gray but tentative and may pass from all true to all false in light of confounding data. Fuzzy logic agrees that scientific truths are tentative but still says they are gray. That is the conflict. Grayness. Whether it be or not be and to what degree. To flesh out this conflict we must first look at the concepts of the philosophy of truth.

THE PHILOSOPHY OF TRUTH:
TRUTH AS A SCORECARD

In modern philosophy the pursuit of truth comes down to how truth behaves, what shadows it casts on the math wall. That starts with two questions: What does truth refer to? What sorts of things are true or false? These questions have the same answer: statements. Truth refers to statements we make, to what we utter or write or nod and point at.

The focus on statements has reduced the analysis of truth to a study of language. Philosophers have stripped truth of its exalted status and reduced it to the way scientists or mathematicians assign 1s and 0s to statements, as if God or an ideal Lie Detector scored our sentences: We get a 1 if what we say is true and a 0 if what we say is false.

Fuzzy theorists have done the same thing. We too see truth as a scorecard but as a gray scorecard: *Fuzzy logic allows more scores*. Fuzzy logic allows the infinite continuum of gray scores between 0 and 1, fits instead of just bits. That is the difference between night and day—at dusk.

STATEMENTS AS VEHICLES OF TRUTH

Statements have truth. What are statements? Statements assert facts or assert states of affairs or assert possible worlds. *Statements assert descriptions*. In practice we write or speak statements as sentences or sentence fragments: Glass breaks. "Glass breaks." "There!"

We say "That makes a statement" when the teenager dyes her hair green or when the executive wears gym shoes or when the movie star wears sunglasses indoors. But here statement means posturing. Assertions make statements in the declarative sense. To make a statement is to assert something. And to assert something is to say it.

Consider a rock. A rock is not true. It just exists. Yet we say that the statement "Rocks are hard" is true. We point or measure or cite a book or expert to show that rocks *are* hard. Then we say that "Rocks are hard" is true since rocks are hard, since it is a fact that rocks are hard.

The statement "Rocks are hard" describes a possible fact, a candidate hypothesis or claim about how the world behaves, how things hang in the space-time continuum, how all the atoms and molecules and quarks and leptons line up in our region of the universe. We say the statement is true if the hypothesis is accurate, if past or present (or future) experience confirms the claim.

Aristotle saw statements as propositions or subject-predicate sentences like "The rabbit runs" or "Jesus wept." Today we use the simpler term "statement." In practice sentences and statements are the same. We get meaning from them and put meaning in them. Sentences are the pipes through which meaning flows.

Statements describe the world. They describe possible worlds. More than that, statements affirm or deny a possible world. Statements combine description with affirmation. A statement is an assertion plus a description. The statement "It rained today" asserts a rainfall event in the space-time continuum. A painting of the same rainfall may "make a statement" in an artistic sense but not in the truth sense, because the painting does not assert or claim or declare or pretend that the rain fell. The artist may claim it, and when she does, she makes a statement. The painting describes but does not assert.

Most modern philosophers are behaviorists. So they define statements as what we affirm or deny. We emit assent or dissent responses when asked a statement. This definition might include paintings and music as statements, so users must tighten how they define assent and dissent or limit how they apply them to spoken or written sentences. The behaviorist approach views thought as silent speech, unmouthed statements, preverbal talk. It views belief, as Bertrand Russell phrased it, as a suspended reaction.

Statements have properties besides truth. They have quantitative properties like length or word count, punctuation, modality, and tense. They also have qualitative or aesthetic properties like irony, clarity, and dialect. The scorecard view of truth reduces truth to just one more quantitative property of statements, like the number of words or commas in a statement.

The scorecard view ignores the meaning of statements. It just scores the sentence with a 0 or 1. If you look at a truth score, you know that the statement is true or false but you need not know what the statement asserts. The statement may be a mile-long formula from differential topology or the radio response from an alien spacecraft. Whatever the statement, a 0 means it is false and a 1 means it is true. The hard part is getting the score right.

LOGICAL TRUTH AND FACTUAL TRUTH

Philosophers distinguish logical truth from factual truth. Logical truth holds for statements that describe logical or mathematical relationships such as "Either A OR not-A" or "1 + 1 = 2." The sentence "1 = 2" is of the logical type, but a logical falsehood, a contradiction. Logical truth comes from symbols and their formal relationships. It does not depend on how the world is or is not.

Factual truth holds for statements that describe possible facts or events, subsets of possible space-time continuums. These include the statements of science and science fiction: "It rained today" and "Man will abolish disease by the year 2400." But "Rain is rain" is not a fact. "Rain is rain" is a logical truth, a tautology, a statement true in all cases, true just by the way the words relate and not by what the words stand for.

Philosophers follow Aristotle and say that factual truths are true in some cases and false in some cases, just as the truth of "It is raining" depends on context. Aristotle called this factual truth *contingency* and called logical truth *necessity*. Logical truths are true in all cases. Logical falsehoods are false in all cases. Contingent or factual truths cover the mixed cases of statements that are sometimes true and sometimes false.

Philosophers often use "possible worlds" to define logical and factual truth. A possible world is a possible space-time continuum, an arrangement of atoms and molecules, a math world. Our space-time continuum, all that we know in space and time, defines one possible world, the actual one. Then a logically true statement is a statement true in all possible worlds. Factual statements are statements true in some possible world, perhaps in only this one. On this view the way we add and multiply numbers should work in all possible worlds. The "laws" of logic and math hold in all worlds. But there may be worlds without radio and TV signals in them because the physical "laws" of electromagnetism need not hold in all worlds.

COHERENCE AND CORRESPONDENCE

The split between logical truth and factual truth grounds two broad theories of truth, the coherence theory and the correspondence theory. Logical truth falls in the *coherence theory* of truth. The statement "$1 + 1 = 2$" is true because we have coherently applied the rules of math. The fairy tale "The Little Mermaid" is true in the sense that it coheres or is internally consistent. It just does not correspond to fact.

Math itself is not true or false. We accept or reject math for practical not logical reasons. Statements *within* a system differ from statements *about* a system. Do numbers exist? The statement "Numbers exist" is true within the mathematical system but makes little sense outside it. We buy the system as a package deal on its merits much as we buy a car or adopt a fad diet or exercise routine. If I hold three rocks in my hand, I need not commit to holding the number three in my hand.

Math is a formal system. We can manipulate math symbols and not understand what they mean. We can just apply the syntax rules as a computer does when it adds up numbers or proves a theorem. The computer shows the truth of the theorem but does not "understand" its "meaning."

All formal systems work this way. You do not have to interpret them to apply them. Propositional logic, with statements like "Socrates is a man," is a formal language. Syntax rules govern its statements or "well-formed formulae." Predicate logic, with statements like "All men are mortal," is a formal language too. The computer languages Ada, Basic, C, Fortran, and Pascal are all formal languages, though they have tended to grow in informal ways to meet the needs of programmers. Statements are true in a formal language if they obey the rules, if they cohere with the rules.

The *correspondence theory* of truth says that a statement is true if it corresponds to fact. "The earth rotates" is true when the earth rotates. It is false when the earth does not rotate—as one day it will not. True statements match or "correspond" to events in space-time. They describe chunks or regions of space-time, the time evolution of whole systems of atoms and molecules. False statements do not describe chunks of space-time or they do not describe the right chunks at the right place-time.

Polish logician Alfred Tarski* modeled this correspondence with his famous *statement formula of truth:*

"STATEMENT" is true if and only if STATEMENT.

The quotation marks mark off an asserted description. The unquoted statement describes a fact. A bilingual example helps make the point: "Grass is green" is true if and only if grass *is* green. "Gras ist grün" is true if and only if grass is green.

Coherence and correspondence reflect math and science, logic and fact, necessity and possibility, syntax and semantics. Philos-

---

*Tarski worked in the early 1920s with Polish logician Jan Lukasiewicz, who first worked out the math of multivalued or fuzzy logic. But like most modern philosophers Tarski had little interest in multivalued logic.

ophers, linguists, and computer scientists have written thousands of papers and books on the logic/fact split.

Coherent and correspondent truth make their own irony. Coherent truth is empty if achieved, and self-contradictory if not achieved—"1 = 1" or "1 = 2." Correspondent truth stands only to fall. Each factual statement is an inaccurate description, an inexact sensory hypothesis that the next experiment may knock down. This inaccuracy of description, where a statement is largely true but not completely true and is thus a little bit nontrue or false, an inaccuracy built into the very "approximate" nature of the scientific method, cracks open the bivalent door and lets the fuzzy logic in.

HEMINGWAY'S CHALLENGE: TRUTH AS ACCURACY

Hemingway said that the goal of a writer is to write one true sentence. He did not mean that in the sense of bivalent factual truth, as modern philosophy and science would interpret it, for then he would have met his goal when as a child he first wrote "My name is Ernest Hemingway." And he did not mean it in the sense of bivalent logical truth, for again he would have met his goal when he first added up a sum of numbers and got the sum right.

Hemingway meant *accuracy of description.* He meant accurate correspondence, accurate match of word with object, theory with fact, finger with moon. He meant the realm of fuzzy factual truth we think in, write in, and run science in.

Fuzzy logic views truth as accuracy. And accuracy is clearly a matter of degree. The precise but artificial statements of math are always 100% accurate or 0% accurate. Statements about the world have accuracy scores between these two extremes. Over the decades fuzzy logicians have developed a vast math machinery that manipulates these accuracy degrees, these fits instead of bits. The fuzzy math always reduces to bivalent black-white math in the extreme case.

Truth as accuracy brings us back to the mismatch problem of gray world with black-white description. Einstein called it right: Logical proof differs from empirical or "scientific" test. If you can prove a statement 100% true, it does not describe the world.

If it describes the world, you can't prove it. You can prove only math things or logic things, coherent things in an arbitrary formal system of made-up rules. That means emptiness. You can prove only tautologies or logical truisms like "Rain is rain" or "A is A" but nothing at all about the real world that science describes. Proof techniques cannot touch the real world. You cannot guarantee what you can test and you cannot test what you can guarantee. Certain math, uncertain world.

Descriptions split into two groups, the logical and the factual, or the mathematical and the scientific, or the coherent and the correspondent. The split depends on accuracy. Logical statements are completely accurate or completely inaccurate. They alone are all or none. Factual statements are partially accurate or partially inaccurate.

The logical and the factual do not meet. This has led some radical empiricist philosophers, such as John Stuart Mill, to see logical or math truths as just the limiting extremes of factual truths just as black and white are the limiting cases of gray. This is a clever idea and a fuzzy one but hard to work with or show. The trouble is the infinite walk to the limiting case. We see only a few finite steps, never a statement that grows before our eyes from partially true to 100% true. There are infinitely many missing links in the fossil evidence.

Ironic as it sounds, inaccuracy is the central assumption of science. No scientific conjecture or hypothesis or theory or statement is 100% accurate. Only the "empty" tautologies of math and logic enjoy that status. Inaccuracy pervades science. The goal of science is to remove as much inaccuracy of description as possible, as much as experimental error and good guesswork and physical "laws" permit.

Scientific claims or statements are inexact and provisional. They depend on dozens of *simplifying* assumptions and on a particular choice of words and symbols and on "all other things being equal." There are just too many molecules involved in a "fact" for a declarative sentence to cover them all. When you speak, you simplify. And when you simplify, you lie.

Even if you could squeeze all the fuzz out of statements, you would still face the problem of all inductive reasoning, the problem that the next measurement may refute your claim. The next link in the causal chain may surprise you. One day the sun will not rise. Or the "law" of gravity may stop being a law and start to change with no apparent cause or pattern. Or in the next moment you may turn into a frog or burst into a fiery supernova or collapse into a black hole. You cannot rig nature to prevent it. You can only take the next measurement.

Scientists get confused on this point because they speak math. They speak accuracy and seek accuracy and reward accuracy. But they achieve only inaccuracy. Scientists try to find the math that best fits the world or the world that best fits the math. They build exact math models to describe some little piece of the universe or to describe the whole thing. They spend their professional time arguing that the few drops of data someone has measured support their models better than they support the competition's models. Or they argue that their math fits with the math of the current champ's model or the math of the competition does not fit. They work in math and trust in math and shoot for math but they can never achieve the logical certainty of math. If scientists minted coins, the coins would bear the logo "In Math We Trust."

Somewhere in the process wishful thinking seems to take over. Scientists start to believe they do math when they do science. This holds to greatest degree in an advanced science like physics or at the theoretical frontier of any science where the claims come as math claims. The first victim is truth. What was inaccurate or fuzzy truth all along gets bumped up a letter grade to the all-or-none status of binary logic. Most scientists draw the line at giving up the tentative status of science. They will concede that it can all go otherwise in the next experiment. But most have crossed the bivalent line by this point and believe that in the next experiment a statement or hypothesis or theory may jump from TRUE to FALSE, from 1 to 0.

At root lies the mismatch problem. In terms of truth the mismatch problem* recounts the antagonism between bivalence and multivalence, between the black and white and the gray:

---

*The mismatch problem has a simple form in the symbols of logic. Let t(S) stand for the truth value or degree of truth of statement S. So t(S) is a percentage. It is a number between zero and one. So it lies in the truth interval [0, 1]—the fuzzy cube of one dimension. Then the mismatch problem says the interval boundary (cube corners) differs from the interior of the interval [0, 1]:

Formal Truth Mismatch:
LOGIC: $t(S) = 0$ OR $t(S) = 1$
FACTS: $0 < t(S) < 1$

The Truth Mismatch

LOGIC: 100% TRUE or 100% FALSE

FACTS: *Partially true* or *partially false*

And never the twain shall meet.

The truth version of the mismatch problem lets us extend Hemingway's challenge* to every scientist in every civilization in every universe ever:

Produce one true factual statement.

And draw a circle if you can.

The Hemingway award goes to whoever finds a statement of fact that is 100% true or 100% false. To win you must pin down which fact your statement refers to. This means you must show the exact match or "correspondence" between words and molecules. Then you must show that the correspondence holds and your fact does what your statement says it does. Then you must tighten this up and show that the fact behaves 100% as your statement claims it behaves—you must show 100% accuracy. This just means you apply the scientific method and state and test a hypothesis. The catch is you must take seriously the bivalent words of scientists. So far no one has met the challenge.

Western philosophers have competed for the Hemingway award since the first days of philosophy. Philosophy is an extreme form of the contest. Philosophy—metaphysics, epistemology, ethics—boils down to making "certain" claims about the world, making logical claims about matters of fact, trying to "prove" nonmath statements. Idealism and empiricism arise from attempts to prove the statement "The world is real" or "The world exists." How do you test that hypothesis? Thought proofs make up the idealist or rationalist category of test procedures: Only ideas are real and I have ideas (Plato). I think and so I am, and so something is, and something is part of everything (Descartes, Spinoza, Leibniz). Touchy-feely finger proofs make up the empiricist or positivist or realist category: I sense something and so something is out there that makes me sense or feel it—maybe nothing exists except my sensations, but at least that is something (Locke, Berkeley, Hume, Russell, Carnap). I move

*Formal challenge: Produce a factual statement S with truth value $t(S) = 0$ or with truth value $t(S) = 1$.

my hand and something wiggles and all the scientists agree (Moore, Quine).

Immanuel Kant based his whole philosophy on a denial of the mismatch problem. He claimed there were "synthetic *a priori*" statements, statements both factually true and logically true. He put forth the math statement "7 + 5 = 12" as an example. He saw the psychological act of counting to seven and then counting to five and then adding the two counts as the synthetic or empirical or factual part of the logical truth.

The logical positivists denied this outright. "There are no synthetic *a priori* statements" was the creed of the Vienna Circle positivists in the 1920s and 1930s. On this point fuzzy theorists agree with positivists. Logic splits from fact. They disagree about what to do about it. Fuzzy theorists take it straight and extend math to fit facts. Black-white math for logic. Gray math for facts. Positivists and scientists and physicians and the like accept the truth mismatch but deny it in practice by making all statements certain, by giving factual statements the status of logic statements. In effect they try to make fact fit their binary math. They call fact statements true or false and mean 100% true or 100% false as if they were logical tautologies or math theorems. And they cover up and cloud the whole issue with probability disclaimers that all such black-white talk holds only with "some probability." Positivists swallow the mismatch whole and use high-pressure sales tactics to get everyone else to swallow it.

I remember when I sat in class and first heard of the positivist Rudolf Carnap, who held sway in philosophy in the 1930s and 1940s and who set up the positivist filter through which all must now pass in philosophy. The professor said Carnap and friends had gotten rid of God, mind, ethics, and metaphysics and reduced philosophy to (binary) symbolic logic and the study of how we use words. Then the professor said, "Men trembled when Carnap walked by." The whole class could tell that Carnap had made the old man tremble too. You can always tell when one man has cowed another. Then the old man said what made me remember it all: "Carnap and the positivists took God out of philosophy. But they put the *fear* of God back in it."

Modern philosophers have dealt with the mismatch problem

by packing the positivist view into Alfred Tarski's statement formula of truth. Tarski put forth "Snow is white" and "Grass is green" as 100% true factual statements—provided snow is white and grass is green. Most modern philosophers agree with him. So they all stand to share equally, and in their view trivially, in the Hemingway award. But not so fast.

What did Tarski mean by his famous example " 'Snow is white' is true if and only if snow is white"? The standard answer is that he meant that this is just what truth is, all or none, 100% or 0%, and he got it right, all right. What would Tarski say about a blade of brown grass? The standard answer is that he would say "Grass is green" is false 100% and thus "Grass is not green" is true 100%. He would flip bits from 1 to 0 or from 0 to 1.

The problem with this lies in tacking on the 100% certainty factors. Tarski had seen as much light-green grass and yellow-green and brown grass and as much gray and yellow snow as you and I have seen. His fine eye saw all the gradations that we see. Tarski may have meant his truth formula in an approximate sense of reasonable men and women doing and saying reasonable things. He developed an elegant but simplified model of truth in a formal system. He rounded off.

The formula itself rounds off. It tacks on the 100% certainty factors by silent default. Or at least philosophers have viewed the formula as including bivalent truth values. But in the same way we can fuzzify Tarski's formula: "Grass is green" is 85% true if and only if 85% of grass is green. Here a game of infinite regress begins. The positivist can make a meta–round off and say this just means that the higher-order statement " 'Grass is green' is 85% true" is true 100% if and only if 85% of grass is green. This admits the existence of fuzzy-truth values, here the 85% truth score. And we can come back and again fuzzify the whole formula. We can fuzzify any attempt to meta–round off our higher-order versions of the formula.

Rounding off lies at the heart of working with symbols and speech. Rounding off compresses information. It simplifies matters and reduces the many to the few, reduces the complex to the manageable. We round off to get by and to get a quick handle on ideas and the pieces of our changing world view. We have to simplify to get things going, at least at first. No harm in that.

But actions have costs. Bivalence or rounding off trades

accuracy for simplicity. When you round off, you pay in truth and accuracy and honesty for what you gain in simplicity and precision and conformity. Denial does not eliminate the costs. A little rounding off, like a little debt, never hurt anyone. But even a little rounding off, like a little bit of pregnancy, can lead to surprises. If you round off too much, you pay the penalty of bivalent self-contradiction and land in paradox. Then bivalent logic ends and fuzzy logic begins and the water glass is both empty and full.

7

The Ways of Paradox

All traditional logic habitually assumes that precise symbols are being employed. It is therefore not applicable to this terrestrial life, but only to an imagined celestial one. The law of excluded middle [A OR not-A] *is true when precise symbols are employed but it is not true when symbols are vague* [fuzzy], *as, in fact, all symbols are.*

BERTRAND RUSSELL
"VAGUENESS," AUSTRALIAN JOURNAL OF PHILOSOPHY, VOLUME I, 1923

It is not surprising that our language should be incapable of describing the processes occurring within atoms, for it was invented to describe the experiences of daily life, and these consist only of processes involving exceedingly large numbers of atoms.

WERNER HEISENBERG
THE PHYSICAL PRINCIPLES OF THE QUANTUM THEORY

The law of contradiction in things, that is, the law of the unity of opposites, is the basic law of materialist dialectics.

MAO TSE-TUNG
ON CONTRADICTION

Touch your mother's toe. Is that incest or not incest? Touch her ankle, her shin, her knee. Is that incest? And so on up.

I first heard that modern version of the ancient Greek "paradox" of logic as a student when I sat in a class on the philosophy of science. It got a laugh but never got an answer. The example seemed to strike the class and the instructor as a contrived anomaly. It struck me as a crack in the ice. Ignoring such paradoxes helped keep fuzzy logic in the closet of black-and-white reasoning for two and a half thousand years.

Two events in the early twentieth century gave birth to fuzzy logic, or "vague logic" as philosophers then called it. First, logician Bertrand Russell rediscovered the classical Greek paradoxes at the foundation of modern math. Then physicist Werner Heisenberg discovered the "uncertainty principle" of quantum physics.

Russell's paradoxes ended thousands of years of blind faith in the certainty of math, bivalent math. Some mathematicians described the effect as "Paradise Lost." Modern mathematicians still struggle to square Russell's paradoxes with their faith in black-and-white math. In the early 1920s Russell laid the logical foundations for fuzzy (vague) logic but never pursued the subject. But he had let the gray cat out of the black-and-white bag.

Heisenberg's quantum uncertainty principle ended, or at least dented, our blind faith in the certainty of science and factual truths. This faith had grown since the days of Isaac Newton to where it largely displaced faith in religion and God. Now science the liberator and way of truth had liberated itself right out of its job of providing truth. Up close it could provide only partial truths, uncertain truths, fuzzy truths. As with the logical paradoxes in mathematics, the uncertainty principles grew in number and extended well beyond physics and will continue to show up in new places in the future. Every time you watch a TV or HDTV screen you see the uncertainty principle of signal processing at work as it squeezes out a tradeoff between the time and frequency parts of the images.

It took the new mathematics of Russell and the new quantum

mechanics of Heisenberg to make us first doubt, really doubt, the logic we inherited from Aristotle. The irony is that Pythagoras's theorem on right triangles, the only math theorem many schoolchildren see, lay at the heart of Russell's paradoxes and Heisenberg's uncertainty principle—and modern fuzzy logic. We will explore the Pythagorean connection later in this chapter. First we examine Russell's paradoxes and Heisenberg's uncertainty principle and see where the Western logic ends and the fuzziness begins.

Sorites Paradoxes: The Road from A to Not-A

The incest sorites paradox is risqué but not contrived. Sorites paradoxes arise everywhere in our word/world schemes because we everywhere negate things and get nonthings. Our words cut the world into a black A and white not-A but the gray world pays no attention. A pink rose defies the rosebud split into red roses and nonred roses.

We can blur the distinction between A and not-A in a simple way. Cut the A thing into little pieces and replace some of the little A pieces with little not-A pieces. Poke a thousand tiny holes in a red rose petal and fill the holes with white petal pieces. Poke enough holes in enough petals and fill them with enough white petal pieces and the red rose shades into a pink rose.

Zeno's original sorites paradox arises this way. Consider a heap of sand. Is it a heap? Yes. Throw out a grain of sand—replace the sand grain with a nongrain. Is it still a heap? Yes. Keep throwing out sand grains and keep asking the bivalent question and eventually you end up with no sand grains and no heap. The heap has passed into nonheap and you can blame no one sand grain. The Greek word *sorites* means a logical chain of statements of the form "If A, then B; If B, then C; If C, . . . If Y, then Z" so that the first term implies the last: If A, then Z. The all-or-none bivalent juice flows from each statement as if it were going down stairs.

Bertrand Russell used a man's head of hair and asked whether the man was bald. Pluck out a hair and ask again. Keep plucking and asking and eventually, after a hundred thousand plucks or so, the man passes from nonbald to bald.

I prefer a cryonics sorites paradox, the first half of which we will all face someday in some way. Is your brain alive (are you alive)? Yes. Kill a brain cell. Is your brain alive? Yes. Keep killing brain cells and eventually you can't keep asking. Now turn it around and thaw a frozen dead brain and apply a smart army of miniature fuzzy robots who are experts in molecular engineering and repair the molecules in your dead brain cell and bring it back to life. Is your brain dead? Yes, still dead. Fix another cell and another, as a mechanic fixes another part and another of a smashed-up car, and eventually your brain lives again and you live again. Something like this happens each morning when we wake and pass from sleep to nonsleep.

These sorites examples may seem too cute and contrived to support the claim that everything involves a sorites paradox. So consider any old thing made of anything. Consider a rock or chair or planet or universe. They are made of molecules. Things and people and places are just sets of molecules, bags of atoms. Some molecules belong to the thing and the rest don't, or so our bivalent language would have us believe. The molecules hovering about the chair's boundary defy classification. But even if we can draw a hard line between chair molecules and nonchair molecules, a sorites paradox emerges. Throw out 5% of the chair's molecules. Is it still a chair? Yes. Throw out 5% more of the molecules. Is it still a chair? Sort of. Keep throwing and asking and eventually the chair passes into a nonchair. The transition is smooth but our description is not. Mismatch again. The old tug of war between gray world and black-and-white language.

Fuzzy logic takes the "paradox" out of the sorites paradox. It comes down to simple arithmetic: Multiply a bunch of certainties and you get certainty. Multiply a bunch of uncertainties and you get compound uncertainty. The more uncertainties you multiply, the more uncertainty you get. Bivalence says the statement "The brain is alive" is true 100%. Fuzzy logic or multivalence says it is true to some degree less than 100%, at first maybe 99% true, then 98% true, and eventually only 1% true when almost all brain cells are dead. To put it in truth terms, bivalence multiplies a bunch of 1s. So it gets back the number one. Fuzzy logic multiplies fractions and gets back a tiny fraction or zero. When all the brain cells are dead, bivalence wants to conclude that "The brain is alive" is still true 100%. Fuzzy logic concludes that the overall chain of implication is vacuous, only 0% true.*

The moral: We pay in certainty points for each inference we make. Reasoning is not a free good. The more steps in our argument, the fuzzier our argument. As we walk down the stairs of inference, of Sherlock Holmes–style deduction, each step becomes less certain, less secure, less persuasive. The longer the explanation, the less we trust it. The best argument is direct evidence or sensory experience. Those are one-step arguments. Nothing persuades like pointing to the balance sheet or the unemployment lines or the cancer boils. All else is guesswork and involves fuzzy inference steps.

We would like to believe we reasoned with Aristotle's logic. That's why Sherlock Holmes and *Star Trek*'s Mr. Spock are heroes and not fictional commoners. We even built our first machines and still build our computers to operate with Aristotle's artificially precise logic. In most cases that never raised machine IQs above the moron level. To raise machine IQs to our meat IQ levels we can make machines reason with our sloppy fuzzy logic, our logic of the Buddha. In terms of the mismatch problem this just means we need gray language to describe a gray world. The sorites paradoxes have been telling us that for almost 3,000 years.

What about mathematical reasoning? It remains bivalent. It, and it alone, reasons in long chains of 100% certainty, long products of 1s. A implies B with certainty, B implies C with certainty, and so on to the certain conclusion. A two-line proof is no tighter than a million-line proof. Mathematicians often judge the "deepness" of a theorem by the number of steps in its proof. In 1976 a computer checked thousands of cases to prove the long-conjectured four-color theorem, which states that you can color a map with only four colors if countries that share

*Bivalence: $1 \times 1 \times 1 \times \ldots \times 1 = 1$. Multivalence: $1 \times .99 \times .98 \times .97 \times \ldots \times .00001$ equals a number very close to zero. The fuzzy product "goes to zero" as the number of uncertainty factors increases to infinity.

borders must have different colors. "Deep theorem" means hard proof, and that usually means long proof. Sorites paradoxes remind us that walking through math differs from walking through the universe.

Fuzzy logic rests on the bivalent reasoning of math. We use lots of little black-and-white bricks to build the mathematical theory of grayness. The question then arises: Can we find a math statement that is gray? The mismatch problem (Hemingway's challenge) makes us give up the search for a statement about the world that is certain, a black-and-white description of a gray thing. But what about the reverse? Can we find a gray description of black-and-white math? Zeno came close. About 2,500 years later Bertrand Russell followed Zeno's lead and found one. He found a math paradox of self-reference.

PARADOX AT ENDPOINTS, RESOLUTION AT MIDPOINTS

Paradoxes of self-reference have the same form. They both assert and deny themselves. They have the logical form of a contradiction, A AND not-A, and they vex mathematicians and Western philosophers. In the fuzzy-principle chapter we saw some examples: the Cretan who says that all Cretans lie, the bumper sticker that reads DON'T TRUST ME, and Russell's barber, the man who shaves all and only those men in the town who do not shave themselves.

There are many more paradoxes of self-reference. All end in A AND not-A:

All rules have exceptions. That's a rule. Does it have exceptions? Suppose it does. Then it is not a rule. But if it has exceptions, then there are rules without exceptions, and it refutes itself. It is both true and false, both A AND not-A. The same holds for the generalization that all generalizations are false.

You can draw your own liar paradox on an index card. On one side write "The sentence on the other side is true." On the other side write "The sentence on the other side is false." Like this:

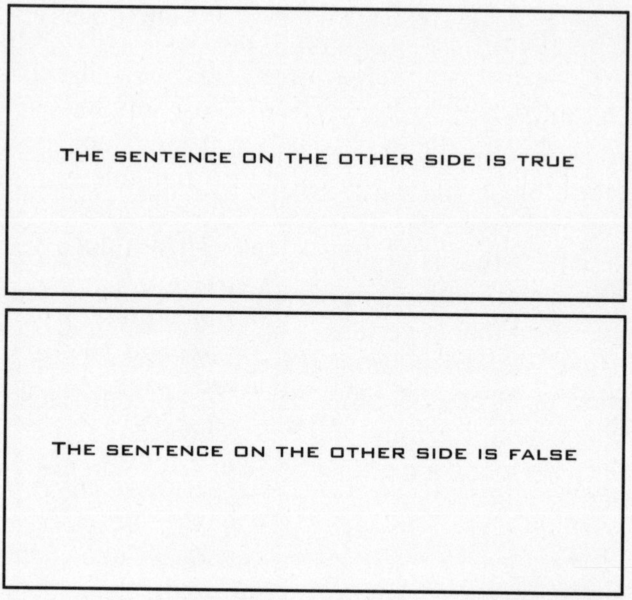

THE SENTENCE ON THE OTHER SIDE IS TRUE

THE SENTENCE ON THE OTHER SIDE IS FALSE

FIGURE 7.1

At this level the paradoxes of self-reference look cute and mere plays on words. That helped give them the name "paradox," which suggests these bivalent contradictions are only apparent problems, that they represent exceptions, that we can fix them with work. Then Bertrand Russell found a paradox that ended the certainty in math that had prevailed since before the time of Aristotle. For this and for many reasons Russell is the grandfather of fuzzy logic.

Russell's paradox deals with sets of sets. Boxes full of boxes. The sets themselves are not fuzzy. Everything is in them or not. Membership is not a matter of degree. Russell found a set where things were in it and not in it. He found *the set of all sets that are not members of themselves*.

Consider the set of apples. Is it a member of itself? No. Its members are apples not sets. The same holds for sets of things like people or stars or universes. They contain people or stars or universes. They do not contain sets. What about the set of all sets? Is it a member of itself? Yes. The set of all sets is a set and so wins membership in its own club. Other sets can also be members of themselves. Infinitely many sets fall in this category. Now what about the set of all sets that, like the set of apples, do not belong to themselves? Most sets we see around us belong to this set. Does it belong to itself? Suppose it does. Then by definition it does not because the sole criterion for membership is that the set be a set that does not belong to itself. So suppose it does not belong to itself. Then it does because it satisfies the membership criterion. A AND not-A.

Russell's set paradox hit the math community as a type of scandal. A few decades before, mathematicians had had to deal with non-Euclidean geometries of curved space. At the turn of the century Georg Cantor made them accept a cascade of infinities, as many infinities as there are counting numbers (we do not know if there are more—I suspect there are more, that there is a continuum of "fuzzy infinities"). But curved space and the ladder of infinities did not challenge the certainty of math. They extended it into new fields. Russell's paradox was not a paradox but a contradiction. That meant you could prove any claim you please since a contradiction implies everything.* And there goes certainty. Russell put math in its first crisis.

The first response was denial. Many mathematicians dismissed the paradoxes as word play. They saw no problems in their branches of math. They found no paradoxes in homological topology or differential geometry or commutative algebra. The paradoxes seemed an artifact of how logicians had cast the foundations of logic and set theory. That attitude continues to this day. Few mathematicians lose sleep over Russell's paradoxes even though the paradoxes affect every branch of math because every branch builds on set theory. The set or collection or class is the fundamental structure in math. In the beginning there were not things. There were sets of things. Even nothing is a set. The empty set.

The next response was to define the paradoxes out of existence. The paradoxes were contradictions. In the proof strategy

---

*Here's how you derive any statement B from the contradiction A AND not-A. The conjunction A AND not-A implies statement A. Statement A implies the disjunction A or B for any B, such as "Dogs fly." (Birds fly. So birds fly or dogs fly.) This disjunction equals the implication IF not-A THEN B. (If birds don't fly, then dogs fly.) The conjunction A AND not-A implies not-A. (So birds don't fly.) Then not-A and the implication together imply B. So dogs fly. Q.E.D.

of *reductio ad absurdum* you make an assumption and show that the assumption leads to a contradiction or "absurdity" and then you go back and deny the assumption. Socrates chopped up sophists with this technique. Politicians use it to attack opponents. We all use some version of it to attack what we don't like. You don't have to argue against the idea or practice. You just have to show that it leads to bad effects: Atheism leads to no morals. Anarchy leads to chaos. Marijuana leads to harder drugs. Pornography leads to rape. In these cases we use the *reductio* technique in simple one-step arguments. If A implies C and if C turns out false or leads the wrong way or tastes bad, then we deny A. Blame the effect on the cause.

The Russell paradoxes were not simple. All of math led to them. Logic led to them. That would be like all photographs causing rape or all food causing heroin addiction. No one knew what math assumption they had to drop to prevent the paradoxes. The argument has more than one step. The effect has several joint logical causes. If A and B imply C and if C turns out false, then either A is false or B is false or both A and B are false. We don't know which. Math rests on several axioms and several intuitions. Which ones caused the trouble?

A search began for a clean set of axioms that both avoided paradoxes and preserved math. Different logicians proposed different axioms of set theory and logic. Russell offered his "theory of types" to avoid paradoxes. This theory said that you could talk down only one level at a time in a hierarchy or math staircase. You could say "Apples are red" because redness is a property of apples and other objects. It hangs just above them in the logical hierarchy. You could also say "Red is a color" for the same reason. But you could not say "Apples are color." That skipped a level. Why couldn't you skip a level? Russell said so. He said so because he thought that would prevent paradoxes. Mathematicians and logicians challenged this assumption and went on to show that Russell's theory of types cost too much math.

The other axiom schemes met the same fate. The new axioms either threw out too much of math or led to new paradoxes. Some did both. They saved math at the expense of math and offered no other virtue. The new axioms did not rest on intuition. They were too arcane and abstract for brains evolved over two hundred million years of mammalian evolution in primitive settings. Language symbols and commerce symbols came late in the process in the last 10,000 years or so—in about the last 500 generations. The only case for a new set of axioms was that it might get rid of the paradoxes with little damage to math.

Russell seems to have been the first to suggest the fuzzy response. He did not pursue it but he made it. Why not drop the law of excluded middle? To hell with Aristotle. Who says A OR not-A must hold for every statement A? But that seemed too radical. No one wanted to give up proof-by-contradiction. Harvard logician Willard Van Orman Quine, the modern successor to Russell, and others said that view was itself a sort of *reductio ad absurdum*.

But the Quine response begged the question. It assumed bivalence when it needed to argue for bivalence. It found multivalence absurd or distasteful just because it was multivalence and not bivalence. And in science as in art there's no accounting for tastes.

The absurd response also missed the key point: Who says all contradictions are the same? Suppose we obey the laws of logic or math and end up with a case of A AND not-A. Who says this statement must hold 100%? The fuzzy view says only that it occurs. We find pink roses that are red and not-red. We find borderline cases just as fast as we define borders.

The fuzzy view says more than this. It says paradoxes of self-reference are *half-truths*. Fuzzy contradictions. A AND not-A holds but A is true only 50% and not-A is true only 50%. The paradoxes are literally half true and half false. They reside at midpoints of fuzzy cubes, equidistant from the black-and-white corners. They correspond to the Buddha who sits grinning at the value ½ at the two Aristotles who sit scowling at the values 0 and 1:

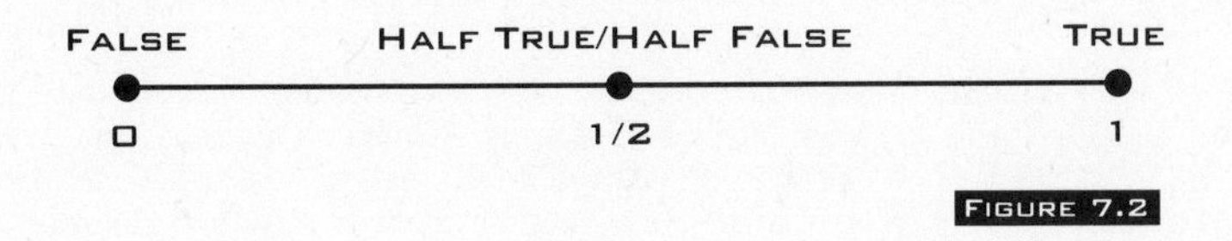

FIGURE 7.2

A simple argument proves that self-reference paradoxes correspond to the midpoint of the truth line from zero to one.* For the most part scientists and mathematicians have not seen this argument. They cut the argument short by cultural fiat when they describe the world with binary logic. But the very scientific method should have tipped them off. An experiment went bad when they found a paradox. Rather than deny the experiment they should have questioned the assumptions behind the experiment. They should have denied bivalence.

Paradoxes generalize to sets as well as logic. Remember that the number line from zero to one defines a hypercube of one dimension. A unit square defines a hypercube or fuzzy cube of two dimensions (with four black-and-white combinations or sets of two objects). A solid cube like an ice cube or a Rubik's cube defines a fuzzy cube of three dimensions (with eight black-and-white combinations of three objects). In each case the corners define the only black-white Aristotelian outcomes where A OR not-A holds. The cube midpoint defines the unique Buddhist outcome where A equals not-A, where the yin-yang equation holds 100%: A = not-A. The midpoint is the only point in the cube that is equally far, and equally close, to every corner. You cannot round off the midpoint to any one corner just as you cannot find a unique direction south when you stand at the north pole of a planet.

The old and new paradoxes teach us many things. The first thing they teach is that we have misnamed them. The term *paradox* suggests exception. A fuzzy analysis shows the reverse. Paradoxes are the rule and not the exception. Pure black-and-white outcomes are the exceptions, the corners of the fuzzy cube filled with gray outcomes. There are two Aristotelian extremes of black and white, 0 and 1, and infinitely many shades of gray between them. A gray shade means A AND not-A holds to some degree.

Paradoxes also show how bivalence costs. You cannot always

---

*Paradoxes of self-reference have the same form: A implies not-A, and not-A implies A. So A and not-A are logically equivalent: A = not-A. The yin-yang equation. So they have the same truth values: t(A) = t(not-A). Here we face a bivalent contradiction of either 0 = 1 or 1 = 0. But suppose we do not insist on binary logic. We know the truth value of not-A equals 1 minus the truth value of A, or t(not-A) = 1 − t(A). So paradox means t(A) = 1 − t(A). We can solve this equation with simple algebra to get t(A) = ½. So the truth lies in the middle.

round off descriptions of fact and get away with it. You trade accuracy for simplicity and you pay for the trade. The sorites paradoxes show that you can conclude nonsense from chains of bivalent arguments, conclude you are not bald after you have plucked out your hair. The self-reference paradoxes show that in the extreme case rounding off blows its own bridge. The bivalent fear of logical contradiction ends in contradiction. Zeno laughs last.

UNCERTAIN WORLD

In the late 1920s Werner Heisenberg stunned the scientific world with his uncertainty principle of quantum mechanics. He showed that you can look closer and see less. Russell had shown that the logic in our minds is uncertain. Now Heisenberg showed that the atoms in our brains are uncertain. Even with total information you could not say some things with 100% certainty.

Heisenberg showed that even in physics the truth of statements is a matter of degree. He made the world face multivalued logic, statements true or false or indeterminate to some degree. He did not work out the math of fuzzy logic. Jan Lukasiewicz in Poland had already done that a decade or so before. Heisenberg made people question bivalent logic. They had taken it for granted for centuries just as Aristotle had taken it for granted. Aristotle and scientists and mathematicians believed that every "well-formed" statement was either true or false. We might not be able to determine the truth of statements about the innards of suns or atoms or aliens on the far side of the universe. But they were knowable. Heisenberg proved that in quantum mechanics some things we can never know. They are unknowable in principle. Heisenberg made doubt scientific. At the time probability theory was the only way known to put this doubt in math form. So rather than shift us from black-white truth to gray truth, the uncertainty principle had the effect of shifting to the probability of all-or-none bivalent truth. In time fuzzy math and fuzzy quantum mechanics may fix that.

You need to know three little-known facts about Heisenberg's uncertainty principle. First, almost everyone gets it wrong. It has nothing to do with how measurement disturbs what you measure, how the thermometer disturbs the soup. Second, there are lots of uncertainty principles and they have nothing to do with quantum mechanics. They have a lot more to do with the signal processing that runs TVs and telephones and your eyes. They are all artifacts of the linear way of looking at the world. Third, uncertainty principles fall right out of the oldest and most important theorem in mathematics, Pythagoras's theorem for right triangles. This theorem in turn falls right out of fuzzy logic and vice versa.

WHAT THE UNCERTAINTY PRINCIPLE SAYS

Most people know quantum mechanics is strange. They do not know the details but they know about Heisenberg's uncertainty principle. They know that Einstein's relativity bends light, digs black holes, slows clocks, and measures the energy of nuclear explosions ($e = mc^2$). And they know quantum mechanics is strange because light comes in quantal packets and behaves both as a particle and a wave. And they know the uncertainty principle sums up the strangeness: You disturb what you measure. They have heard about it in school or in movies or at a party. Pop-science writers cite it when they write about the history of science or the detection of subatomic particles. Journalists and social scientists point to it to illustrate how the news media disturbs the news. Parents tell it to their kids when the kids ask them about atoms or microprocessors or free will.

And they all get it wrong. Pollsters should test this and let us know how deep the embarrassment runs. Here truth is stranger than fiction. There is nothing strange about a thermometer that disturbs the soup. The fiction lies in that stove-top description of the uncertainty principle.

The truth is strange: If you drive very fast down a freeway in a straight line and watch your speedometer, then you may not know where your car sits on the freeway at any moment. If you drive fast enough, faster than our cars and gravity allow, then you *cannot* know how fast you drive if you know where your car sits on the freeway, or you cannot know where your car sits if you know how fast you drive. Speed or position. You get one or the other. At slower speeds you face tradeoffs. The more you can pin down your speed (velocity) the less you can pin down your position and vice versa. Here speedometers and thermometers disturb nothing. They measure only the effects of the cause. And the cause lies in the nature of things. In the math of things. In the *linear* math we attribute to things.

We drive our cars and fly our airplanes at speeds so slow that the Heisenberg tradeoff is negligible. We can know our speed and position up to several decimal places of accuracy. Evolution and daily experience have formed our intuitions around these slow velocities. So we took millions of years to grasp Heisenberg's world of quantum mechanics as well as Einstein's world of special relativity. No one even guessed at them. The speed (momentum) uncertainty in a drifting dust particle equals about 10^{-26} or less than a trillionth of a trillionth of a unit of velocity (momentum).

You can picture Heisenberg's uncertainty principle as a relation between two graphs. One graph shows our knowledge of the car's position. The other graph shows our knowledge of the car's velocity. See Figure 7.3.

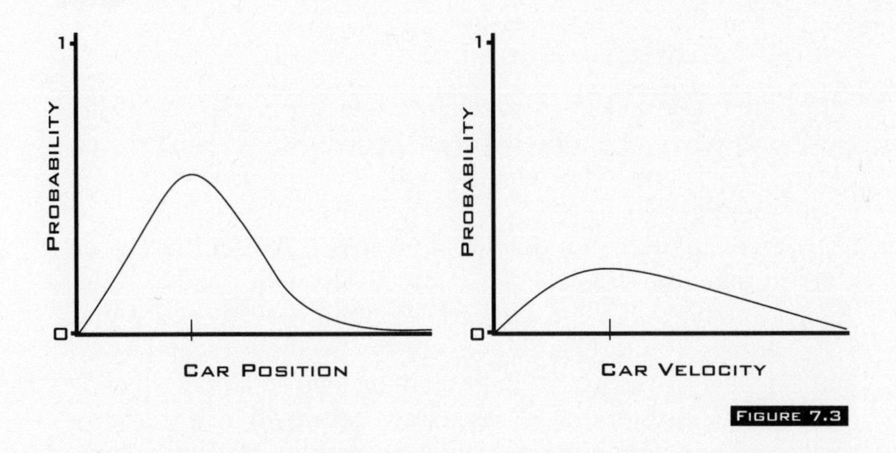

FIGURE 7.3

A bell curve shows the "spread" or variance in our knowledge or certainty. The wider the bell the less we know. An infinitely wide bell is a flat line. Then we know nothing. The value of the quantity, position, or speed could lie anywhere on the axis. An infinitely narrow bell is a spike that is infinitely tall. Then we have complete knowledge of the value of the quantity. The uncertainty principle says that as one bell curve gets wider the other gets thinner.* As one curve peaks the other spreads. So if the position bell curve becomes a spike and we have total knowledge of position, then the speed bell curve goes flat and we have total uncertainty (infinite variance) of speed.

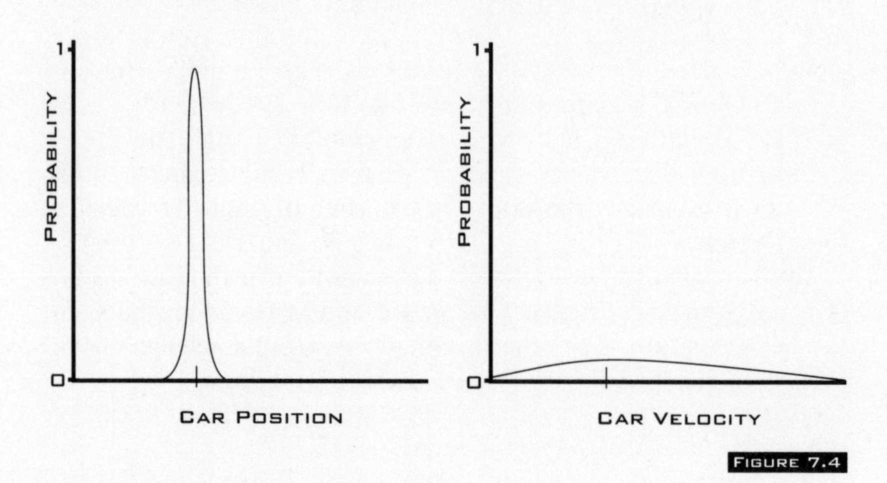

FIGURE 7.4

Heisenberg found this uncertainty relation in the late 1920s. He found other relations between "conjugate" variables, between energy and time, electric and magnetic field strengths. Quantum theorists soon saw that uncertainty relations arose between many quantum operators or objects. (Things are operators in quantum mechanics.) It had to do with operators that did not "commute." The numbers we multiply always commute: 3 × 2 equals 2 × 3. Quantum operators need not commute and that can give rise to uncertainty relations between operators.

Many scientists believed, as many still do, that uncertainty relations were unique to quantum mechanics. But uncertainty relations arose from a math quirk. Many other fields use the same math and create their own uncertainty relations. We have

---

*Let p denote the mathematical variance or dispersion of position and let v denote the variance of velocity. Then the Heisenberg uncertainty principle is an inequality: $p \times v \geq \hbar$. $\hbar$ denotes Planck's constant h (6.626×10^{-34} joule seconds) divided by 2π. Planck's constant is small but not zero. So p and v cannot be zero except as a limiting case. If p decreases to zero, then v increases to infinity since $v \geq \hbar / p$.

built uncertainty relations into the structure of the information age. Natural selection seems to have built them into our brains. It all has to do with right triangles. But we must make one more stop before we get there. We must take a second look at that "strange" theory of quantum mechanics and make another distinction, this time between lines and nonlines, between flat surfaces and bumpy surfaces.

ANOTHER MISMATCH: LINEAR MATH, NONLINEAR WORLD

Don't let quantum mechanics scare you. It's not "strange" at all. *Quantum mechanics is linear.* It's as linear a theory as we have. It's so linear and the world is so nonlinear that many of us have little faith in quantum mechanics except as a rough first cut at all the nonlinear reality that swirls around us and in us. Werner Heisenberg does not win the Hemingway prize for producing the first 100% true factual statement. The uncertainty principle is only an approximation.

Up close you can approximate any wiggling curve with a straight line. And up close the round earth looks flat.*

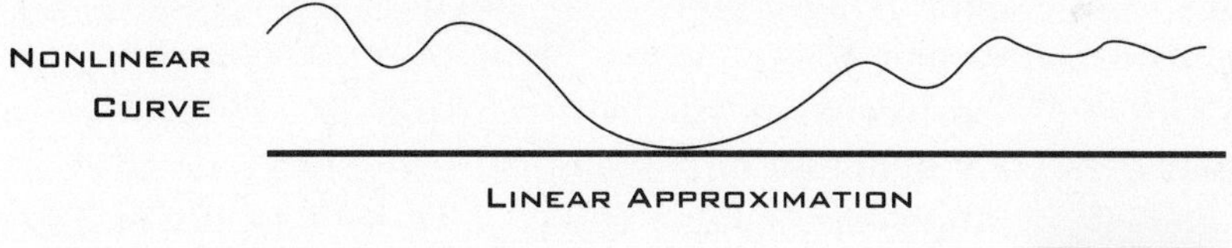

FIGURE 7.5

*Consider Einstein's mass-energy equation, $e = mc^2$. Here c denotes the constant speed of light in a vacuum. Did you know this is a linear theory? The term c^2 is a constant and not a variable. It is just a constant of proportionality. So the energy-mass relationship is *linear*. Increase the mass tenfold and you increase the energy tenfold and vice versa. Did Einstein really believe nature was so linear? No. Even many physicists forget that Einstein derived this linear equation as an approximation in his 1905 paper, "Does the Inertia of a Body Depend Upon Its Energy-Content?" Einstein linearized the math to simplify the math. He ignored infinitely many nonlinear or "higher-order" terms on one side of the equation. Einstein arrived at the lone linear term by "neglecting magnitudes of fourth and higher orders." The real equation reads

$$e = mc^2 + \textit{infinitely many terms}.$$

Scientists and engineers make this sweeping assumption all the time and have a name for it—*linearization*.

In the same way, you can view a linear system as a sheet of paper and a nonlinear system as a crumpled sheet of paper. No matter how bumpy the crumpled sheet is up close, in a small or "local" region the crumpled sheet looks like a flat sheet of paper. But when you stand back and get a "global" view you see that the two pieces of paper differ.

A linear theory gives you the whole from the parts. Add up the parts and you get the whole. Study how the parts behave. Then stitch the parts together and you have studied the whole. Quantum mechanics does this with matter waves and light waves. You can add up waves to get one big wave and you can decompose a big wave into several small waves. Mathematicians call it "superimposition." The uncertainty principle is part of this linear package.

A nonlinear theory does not give you the whole from the parts. The parts do not add up to the whole. That is nonlinearity. Groups do not behave as their members behave. You can study arms and legs and organs and other parts and still not know how a human behaves or how a mob behaves. System complexity exceeds subsystem complexity. That still baffles atomic physicists who use big chunks of matter to study and puzzle over the finest divisions of matter. It baffles medical students who cut up cadavers and wonder where the motion and emotion went. It baffles econometricians who try to predict employment and interest rates with simple linear models of an economy. As Bertrand Russell said, every man has a mother but mankind does not have a mother.

Scientists try to make things simple. That is in good part why we are stuck with bivalence. Scientists' first instinct is to fit a linear model to a nonlinear world. This creates another mismatch problem, *the math modeler's dilemma: linear math, nonlinear world.*

We know a great deal about linear math. In comparison we know almost nothing about nonlinear math—except that almost all math is nonlinear. Mathematical research will never stop. In a finite universe physics and chemistry and biology and economics can stop. Math never stops. The sea of math is infinitely vast. And no matter how much we explore of it, even if we work at it for an eternity, we will never know more than an infinitesimal point in the sea of nonlinear math.

So it is a good bet that someday quantum mechanics will fall because it is linear to its core. It says there are not things in the universe but linear "operators" awash in the cosmic ocean of matter waves. And we observe only the average footprints (the "eigenvalues") of these linear operators like energy values. A nonlinear theory will no doubt replace that. The nonlinear theory will better fit the facts. Quantum chaos and its frothing "random"-looking states of subatomic equilibrium offers the first step in the nonlinear direction. For now we are stuck with a linear quantum mechanics and the uncertainty relations it spawns.

Heisenberg's uncertainty principle is part of the linear model. A nonlinear quantum mechanics may not lead to an uncertainty principle. And in the class of linear systems, which dominate modern engineering, uncertainty principles are commonplace.

The popular-science press has so tied the Heisenberg uncertainty principle to quantum physics that I want to digress to show that uncertainty principles are not special. The popular view not only gives the wrong description of uncertainty principles as measurement disturbances but misses the linear "cause" of uncertainty principles.

Where you have linear theories you will have uncertainty relations. Electrical engineers found these long ago because they too work with linear systems. Linear navigation and guidance systems put men on the moon and bring them back home. Linear systems control microwave ovens and antennae and high-speed modems. In signal processing they cancel echoes in long-distance telephone calls and cut down on the interference in television and radio and radar receivers. Here the linear system is the LTI system, the linear *time-invariant* system. Time invariance means the system structure does not change too fast.

Math Fact: *Every LTI system has an uncertainty principle*. The popular-science press never told you that. Even most physicists do not know it.

The LTI uncertainty principle says you cannot pin down a signal completely in time and in frequency. Signals have time duration. Bird songs. Whale songs. Pop songs. Seismic readings. Blood pressure. TV signals. We sense these signals as intensities that vary with time. They strike our surface receptors and measurement devices. Sound strikes our cochleas and light intensity strikes our retinas.

These signals also have a frequency component called the *frequency signal*. Spectral analysis resolves the bird song into a series of frequencies. All the frequencies are present in the time signal but present to different degrees. A Fourier transform converts the time signal into the frequency signal just as a prism splits white light into beams of colored light.

The uncertainty relation holds between the variance or spread of the time signal and the spread of the frequency signal.* The clearer the time signal the less clear the frequency signal. Dennis Gabor, the inventor of the hologram, proved and popularized this time-frequency uncertainty principle in his famous 1944 paper "Theory of Communication." Two decades before at Bell Labs H. Nyquist and R.V.L. Hartley each arrived at the same time-frequency uncertainty principle but with less clear mathematical arguments. Hartley stated the uncertainty principle in communication terms as a tradeoff between signal frequency and transmission time: "The total amount of information which may be transmitted is proportional to the product of the frequency range which is transmitted and the time which is available for the transmission." Their work preceded Heisenberg's work by a year or so. Neuroscientists have used Gabor's work to show that brain cells in the visual cortex of kittens, and perhaps in our own, fire in a way that minimizes the LTI uncertainty. It all rests on the Pythagorean Theorem. And so does fuzzy theory.

PYTHAGORAS AND THE UNCERTAINTY PRINCIPLE

The Pythagorean Theorem on right triangles is the most important theorem in mathematics. Pythagoras was born around 580 B.C. and claimed that "all things are numbers." He first proved the theorem in the sixth century B.C. in southern Italy. Today it

*Let s denote the variance in the time signal and let S denote the variance in its frequency signal or Fourier transform. Then the LTI uncertainty principle states an inequality:

$$s \times S \geq 1/16\pi^2.$$

For details and a classic debunking of the physics monopoly on uncertainty principles see *Digital Filters*, third edition, Prentice Hall, 1988, by information-theory pioneer Richard Hamming.

lies at the heart of what we call Hilbert space, which in turn lies at the heart of quantum physics and modern engineering.

The Pythagorean Theorem underlies our modern notion of "optimality," the best possible solutions to problems. What is the best prediction of how fast the AIDS retrovirus spreads given all data to date? What is the best way to fill in missing bits in a transmitted message of 1s and 0s or missing gaps in a speech signal or missing pieces of a TV image?

In math and engineering the optimal answers often arise from *orthogonality conditions* that describe when two abstract objects intersect at a right angle as if they were the legs of a right triangle. The orthogonality conditions show us the best way to draw a trend line through a cloud of data points or the best way to filter noise from your car radio or the best way to predict where to aim the SAM warhead when the enemy jet flies behind a cloud.

Pythagoras's theorem rests on the right or 90° angle of the right triangle. The two perpendicular legs of the right triangle have lengths a and b and the hypotenuse has length c:

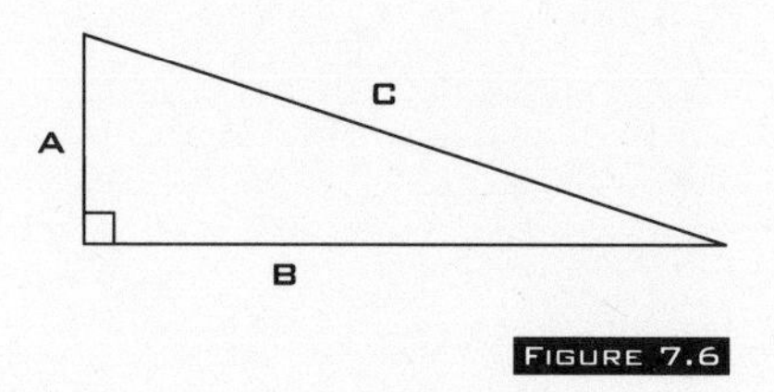

FIGURE 7.6

The Pythagorean Theorem relates the squared lengths of the triangle.*

What do squares have to do with a right triangle? Schoolteachers sometimes forget to point this out when they first teach the

*The Pythagorean Theorem states $c^2 = a^2 + b^2$. Here the square c^2 just means c multiplied by itself ($c^2 = c \times c$) and similarly for the squares a^2 and b^2. The two most famous examples are the 3-4-5 right triangle, since $5^2 = 25 = 9 + 16 = 3^2 + 4^2$, and the two right triangles that arise when a diagonal bisects a square of side length 1: $2 = 1^2 + 1^2$. The last example shocked the Pythagoreans because it meant the square's diagonal (the right triangle's hypotenuse) had the "irrational" length $\sqrt{2}$, a number they first proved could not equal a ratio of whole numbers.

Pythagorean Theorem to children. The Pythagorean Theorem really describes the area of three squares that abut the sides of the a-b-c right traingle.

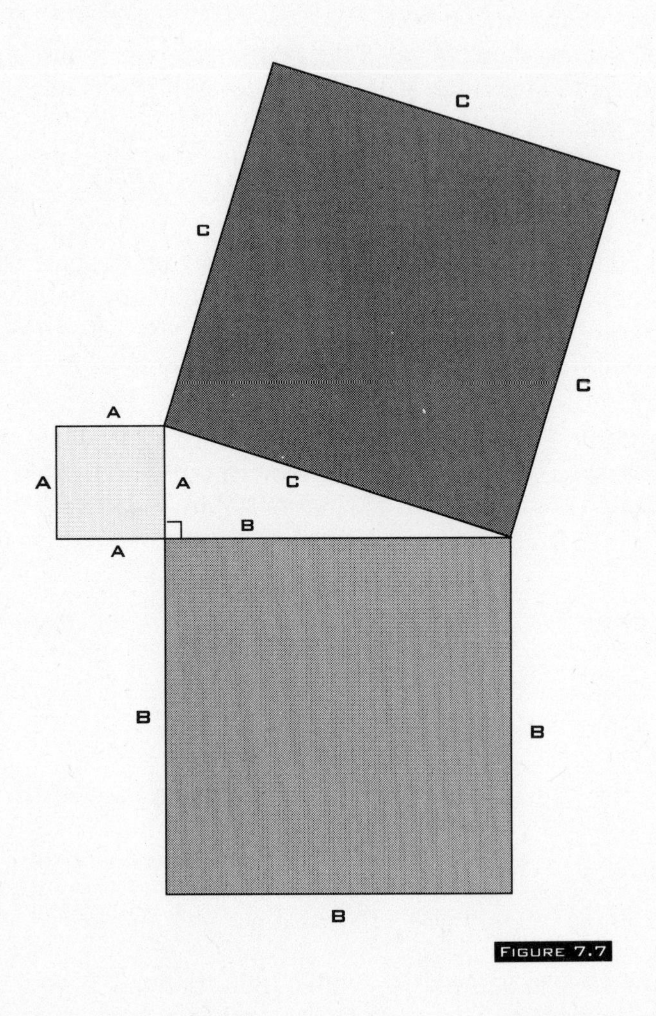

FIGURE 7.7

The Pythagorean Theorem says that the area of the largest square equals the area of both of the smaller squares.

The Pythagorean Theorem arises everywhere in math. Whenever you pick two abstract objects in an abstract "space" a right-triangle relationship holds between them. Consider two points A and B in the plane (Figure 7.8).

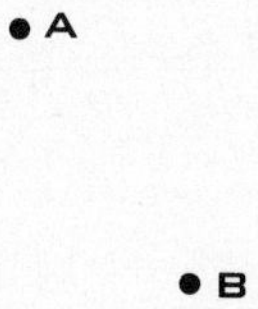

• A

• B

FIGURE 7.8

Points A and B are simple abstract objects. Suppose A and B define the tips of two arrows (or "vectors") drawn from the same starting point.

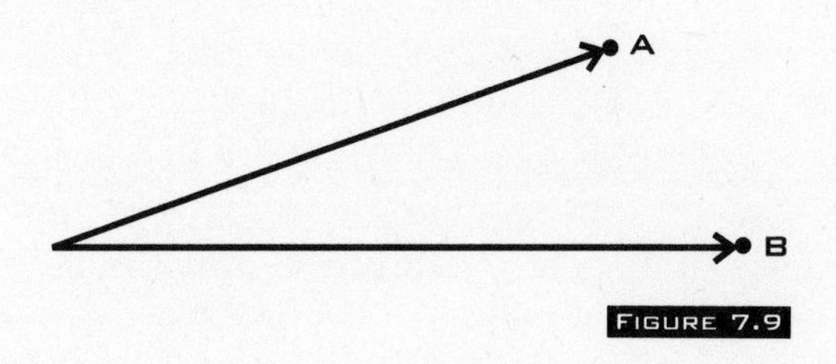

FIGURE 7.9

So where's the Pythagorean relationship? It arises from an "orthogonal" or perpendicular *projection* of the A line onto the B line. You can think of the projection as the shadow the A line casts on the B line when a flashlight shines straight down on the A line from the top of the page (Figure 7.10).

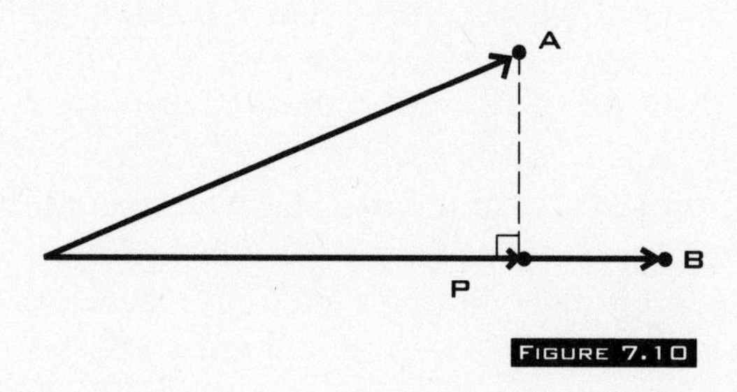

FIGURE 7.10

The A-P-B triangle is a right triangle. So the Pythagorean Theorem applies: We can draw the three squares that abut the triangle's sides and the area of the largest square equals the sum of the area of the two smaller squares. The point P leads to one of the most important sets of equations in science and engineering, the so-called "normal equations" of least-squares curve fitting. The normal equations show how to draw the best-fitting curve through a cloud of data points, a technique Karl Friedrich Gauss first developed in 1795 to predict the return of Halley's Comet from astronomical measurements.

The A-P-B triangle contains the Heisenberg uncertainty principle and the Subsethood Theorem of fuzziness discussed earlier in the chapter "The Whole in the Part." The uncertainty principle applies the *Cauchy-Schwarz inequality* that relates the lengths of the A and B arrows. An inequality, or less-than sign "<," holds when A and B are not parallel, when the A-P-B right triangle has an A-P leg. The inequality *is* the uncertainty principle. Equality holds when A and B are parallel, when arrow A equals arrow P.* The equality case defines the minimum-uncertainty case that seems to describe how the neurons fire in the visual cortex in our brains.

I found it a great letdown when I first learned that Heisenberg's uncertainty principle was just the well-known Cauchy-Schwarz inequality in physics disguise. Every U.S. college kid in science learns that formula in her second or third course in calculus. Kids in other countries learn it in high school, as Heisenberg did. It was a letdown and yet it was an insight. It meant unity. The many under the one. It took me some time before I saw the Pythagorean connection. That was a bigger letdown and bigger insight and sense of unity.

It was years later when I found that you could generalize the math of the Pythagorean Theorem and that it gave you the subsethood theorem of fuzziness, the measure of the whole in the part, and that it swallowed up the old notion of "randomness" or the probability of a part. The waterbed in the sky. Tie a string to the closest point on the waterbed. Tie that string and you have made an orthogonal projection. You have defined an

---

*The Cauchy-Schwarz inequality relates the lengths l_A and l_B of the A and B arrows to the absolute value of the "inner product" or correlation between A and B:

$$l_A \times l_B \geq |A \bullet B|$$

All uncertainty principles have this form.

A-P-B right triangle and the Pythagorean Theorem holds. Or shove a dictionary in the corner of a room and drop a marble on it. The marble falls at a right angle to the surface of the dictionary. Its path defines one leg of the A-P-B right triangle. The whole ensemble of square room, dictionary in the corner, and falling marble describes the geometry of the subsethood theorem of fuzziness.

The fuzzy past might have been truly fuzzy if Pythagoras had made this connection almost 3,000 years ago. But he still struggled with numbers and did not even consider sets of things. The math of sets is barely a century old. Instead in the West we inherited from Pythagoras and the ancient Greeks a belief in absolutes and pure forms, in the white and the black, the true and the false.

It could have turned out otherwise but it did not. Maybe it has in other civilizations in other parts of the universe. There are trillions of stars growing along their "main sequence" profiles and passing through billion-year periods when they might support water on orbiting planets and planetoids.

Question: Of all civilizations who have learned to count, how many have discovered fuzzy mathematics? How many have stumbled on something as intuitive and expressive as a fuzzy set, a group of elements that belong to only some degree to the group as a man belongs to only some degree to a political party or to a circle of friends or to an economic class? I bet at least half have found fuzzy logic. The paradoxes are too common and the Pythagorean math too simple for them to miss it for long.

Even we here on Earth have found it and in the West at that. The culture of ancient Greece was a statistical fluke. Pythagoras or Plato or Aristotle might have died young. The Hittites or Assyrians or Egyptians might have been luckier and conquered the European tribes and brought them the cultures from the East along the Silk Road. Even the rise of bivalent science was a statistical fluke. All that matters in science is the math that tests bear out and you can come upon that from many directions and from many points of view. Don't expect the "scientists" in the Magellan Cloud to wear white smocks and to advise governments and to bully culture and fashion. They may have forgotten so much of their past that they view the method and tools of their science as common sense or something beneath sense and interest. How many of us can chip a flint rock to a sharp point and edge and lash it to a straightened stick or can hunt and quarter and smoke game or can tan a hide or plank a ship or build an arch?

Fuzziness has arrived even if most scientists and engineers do not know it. Thank the Buddha and thank Zeno and thank Russell and Heisenberg and all the rest. The seed took. We now turn to what we have done with fuzzy logic in the fuzzy present.

III

The Fuzzy Present

8

THE FUZZY PRESENT

It is usual with mathematicians to pretend that those ideas, which are their objects, are of so refined and spiritual a nature that they must be comprehended by a pure and intellectual view. The same notion runs through most of philosophy. But to destroy this artifice we need but reflect that all ideas are copied from our [sense] impressions.

DAVID HUME
A TREATISE OF HUMAN NATURE

How do fuzzy systems work? How can shades of gray raise the machine IQ of cameras and vacuum sweepers and helicopters and cruise missiles?

This part looks at these questions in three chapters. The first chapter looks at fuzzy sets and their history. Fuzzy sets are the building blocks of all fuzzy systems. Small numbers. Warm air. Tall men. Most of our words stand for fuzzy sets. And most of modern science still stands against the logic of fuzzy sets.

The second chapter shows how we reason with fuzzy sets in fuzzy systems. If the air is cool, then turn down the air conditioner. If the water is dirty, then add more detergent. If the center of the image is blurry, then turn the lens a little. All the new fuzzy products follow the same design. You can see how a fuzzy system works without any math. You come up with fuzzy rules and put them in the system. We do that for a simple fuzzy air conditioner. When you ask the system a question or give it an input it "fires" or activates all its rules to some degree. It then averages the outputs to give the answer or final output. The scheme is simple and you can describe its geometry with a cookie cutter.

Today there are several cheap fuzzy software tool kits on the market. After you read this chapter you will know how to use these tool kits to build your own fuzzy systems. You will also have some feel for the economic and political context of fuzzy systems in Japan.

The third chapter looks at adaptive fuzzy systems. These systems learn from experience. Neural nets combine with fuzzy systems to help the fuzzy systems grow and tune their own rules. This third part also looks at fuzzy cognitive maps and at how you can use them to model social dynamics and to turn a politician's speech into something you can test.

Everyone should know how a fuzzy washing machine works.

9

Fuzzy Sets

Everything is vague to a degree you do not realize till you have tried to make it precise.

Bertrand Russell
The Philosophy of Logical Atomism

Think of arm chairs and reading chairs and dining-room chairs, and kitchen chairs, chairs that pass into benches, chairs that cross the boundary and become settees, dentist's chairs, thrones, opera stalls, seats of all sorts, those miraculous fungoid growths that cumber the floor of the arts and crafts exhibitions, and you will see what a lax bundle in fact is this simple straightforward term. I would undertake to defeat any definition of chair or chairishness that you gave me.

Charles Pierce
Dictionary of Philosophy and Psychology, Volume 2, 1902

The vagueness of the word chair *is typical of all terms whose application involves the use of the senses. In all such cases "borderline cases" or "doubtful objects" are easily found to which we are unable to say either that the class name does or does not apply.*

Max Black
"Vagueness: An Exercise in Logical Analysis," Philosophy of Science, Volume 4, 1937

Classical logic is like a person who comes to a party dressed in a black suit, a white, starched shirt, a black tie, shiny shoes, and so forth. And fuzzy logic is a little bit like a person dressed informally, in jeans, tee shirt, and sneakers. In the past, this informal dress wouldn't have been acceptable. Today, it's the other way around.

LOTFI A. ZADEH
COMMUNICATIONS OF THE ASSOCIATION FOR COMPUTING MACHINERY, VOLUME 27, 1984

Words stand for sets. The word *house* stands for many houses. It stands for different houses for each of us because we each have seen and lived in and read about and dreamed about different houses. We all speak and write the same words but we do not think the same words. Words are public but the sets we learn are private. And we think in sets.

House stands for a set of houses, a list of houses, a group or collection of things, and each thing we can point to and call "house." But which structures are houses and which are not? You can point out some things as houses more easily than you can point out others. What about castles and trailers and mobile homes and duplexes and time-share condos and teepees and yurts and lean-tos and caves and tents and cardboard boxes in alleys? It's a matter of degree. Some structures are more "a house" than others are. They are to some degree a house and not a house. Exceptions blur the boundary between house and nonhouse. So A AND not-A holds. So fuzziness holds: The noun *house* stands for a fuzzy set of houses.

It does not stop with nouns. Add an adjective and you still get a fuzzy set. *Old house* stands for a smaller set of houses. It stands for a subset of our set of houses. Every old house is a house but not every house is old. But how old is old? Some houses are older than others. It's a matter of degree. Some old houses belong more to the set of old houses than others belong. We can even rank them by year or month or minute. *Very old house* stands for an even smaller set of houses, a subset of our set of old houses. We can also rank their very-oldness by their age. The historical medieval cottage is more very-old than the Quaker farmhouse or the Civil War log cabin.

Think of that: Our brains are full of fuzzy sets. We think in fuzzy sets and we each define our fuzzy boundaries in different ways and with different examples. We stack the furniture of the universe into fuzzy sets. We group things into loose fuzzy sets and then play with the groups and look for connections. *Thought is set play*. That is just what fuzzy logic is—reasoning with fuzzy sets.

Fuzzy logic makes our computers think in fuzzy sets. And that's the advance—getting computers to reason with fuzzy sets. Not words or symbol lists or "language strings," as the computer scientists call them, but fuzzy sets.

This chapter explores the world of fuzzy sets. We explore its ideas and math and history and politics and how it paves the way to the fuzzy systems that raise the machine IQs of electric shavers and microwave ovens and robot arms on space shuttles. I want to show that the fuzzy set is not a fad idea or computer gadget or exception to somebody's black-and-white rule. The fuzzy set is a hallmark of human and machine intelligence, the pure wedding of symbol and idea, the way we cope with a gray world. The fuzzy set is *expressive*.

NUMBERS ARE FUZZY TOO

Everything is a matter of degree. So what about numbers? Numbers are the hallmark of precision. Can we fuzz these up too? As Kant might ask: Are fuzzy numbers possible?

The old answer was no. How could numbers be fuzzy? Numbers are "pure forms" and they either are or aren't. Numbers belong all or none to sets of numbers. Odd numbers. Even numbers. Every whole number is either even or odd, no in between, no middle ground, no gray cases. Scientists have devised a mathematics that seems to have escaped the sloppiness of our everyday fuzzy sets and fuzzy thoughts. Recently they have gone further and tried to keep fuzzy sets out of computers. And they have succeeded—and computers are dumb. Computers add numbers well but they can't recognize a face or a house in most images. Where could fuzziness hide in the black-and-white world of math?

Consider the number zero. 0. It took some ancient societies thousands of years to find zero. Now we can find it at a glance as a spike on a number line (Figure 9.1).

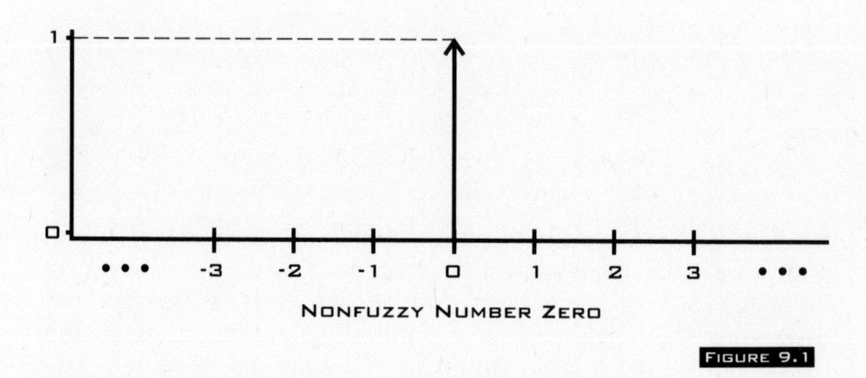

FIGURE 9.1

The spike means the number zero belongs 100% to the set ZERO and no other number belongs to it. Every number is either in the ZERO set or out of it. All or none. In this set sense the number zero alone belongs to the set ZERO.

But what about the numbers *close* to zero or *almost* zero or *nearly* zero? These numbers, like big numbers or medium numbers or very small numbers, are fuzzy numbers. They define a spectrum of numbers near zero and some belong more in the set than others belong. The closer a small number to zero, the more it belongs to the fuzzy set of small numbers. The number 1 is closer to 0 than the number 2 is, and 2 is closer than 3 is, and so on. Likewise the negative number -1 is closer to 0 than -2 is, and -2 is closer than -3 is, and so on. The number 0 belongs 100% to the set ZERO but close numbers may belong only 80% or 50% or 10%. We might draw the fuzzy number zero as a bell curve or triangle centered at the exact number 0 (Figure 9.2).

If we draw the triangle narrow enough, we get back the spike of classical mathematics. That's another surprise: Math as we know it is but a special case of fuzzy math, a special limiting case—the degenerate case of black-and-white extremes in a math world of grays. We can add and subtract triangles just as we add and subtract spikes (numbers). We can also draw the fuzzy number ZERO in infinitely many ways. Each person can

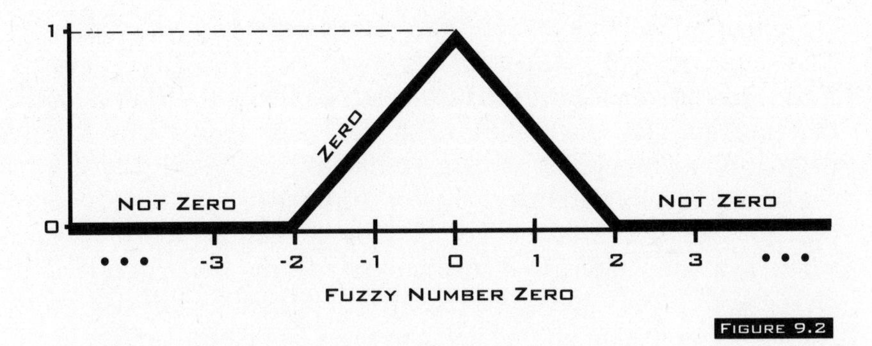

FIGURE 9.2

draw it differently just as each person thinks differently of HOUSE or SMALL or SMART or FAIR or NICE or CLEAN. There are as many ways to draw the fuzzy number ZERO as there are numbers.

What can you *do* with fuzzy numbers? You can reason with them. We do it all the time. "Fuzzy logic" means reasoning with fuzzy numbers and fuzzy sets. In practice it means making computers reason with fuzzy numbers in the form of if-then statements or rules of thumb: If the car goes TOO FAST, then SLOW down (raise the gas pedal SOMEWHAT). If the car backs up TOO MUCH to the left, then turn the steering wheel A LITTLE to the right. If the car traffic is HEAVY in the north-south direction, then keep the traffic light green LONGER. Each term stands for a fuzzy set. Every roadful of traffic is heavy to some degree. Every green light stays on long to some degree. The knowledge or intelligence comes from associating these two fuzzy events.

For now we have proved the point: Numbers are fuzzy too. And we work with fuzzy numbers all the time. And if numbers are fuzzy, everything is. And everything is.

The next question is what is special about fuzzy sets. Do they lead to anything new? Fuzzy sets arise when borders blur, when thing overlaps nonthing, when A overlaps not-A. Does this lead anywhere special? The state of overlap or contradiction A AND not-A cannot hold in bivalent logic. So scientists might have missed some property here. And they have. They missed fuzzy entropy (just as they missed subsethood and mistook it for an undefined "randomness" or "probability"). If there are no fuzzy sets, then there are no properties of fuzzy sets. But there are fuzzy sets and the first question is just how fuzzy they are. How do you measure the fuzziness of a fuzzy set?

FITS OF FUZZY ENTROPY

Fuzzy entropy measures the fuzziness of a fuzzy set. It answers the question How fuzzy is a fuzzy set? And it is a matter of degree. Some fuzzy sets are fuzzier than others.

Entropy means the uncertainty or disorder in a system. A set describes a system or collection of things. When the set is fuzzy, when elements belong to it to some degree, the set is uncertain or vague to some degree. Fuzzy entropy measures this degree. And it is simple enough that you can see it in a picture of a cube. First some recent history.

I found the measure of fuzzy entropy in 1985 while I was first looking for the geometry of fuzzy sets, first playing with Rubik's cubes in my office and then later presenting the results as lectures and homework assignments in my fuzzy class at UC San Diego. I wrote papers on fuzzy entropy and stuck it in my Ph.D. dissertation and came back to it many times after that. Each time I saw a little more of the geometry of fuzzy entropy. Then years later in 1991 I found a link between fuzziness and the classical entropy of thermodynamics and information theory. You can even reduce addition to a general or "mutual" fuzzy entropy, but that is another story. In early 1986 I had proved the fuzzy entropy theorem that says fuzziness equals a ratio of the Buddha over Aristotle, a balance of the Buddha's A AND not-A and Aristotle's A OR not-A. You'll see this below in a moment.

In the first chapter I showed how people in an audience can give a fuzzy set when they raise their hands to answer a question. Are you married? Raise your hands if you are. That separates the audience into two pieces, into an A piece and a not-A piece, the married and the not married. The law draws a hard line between married and not married. Now put down your hands. How many of you are happy? Or young? Or rested? Or liberal? Or thin? Or tall? Or smart? Or honest? Raise your hands. Now the hands do not go up all the way or stay down all the way. No law draws a fake line for us between happy and not happy, young and not young, honest and not honest. Our fuzzy logic does not draw hard lines between opposites. We live with a mix of happiness and unhappiness, fit and fat, honesty and dishonesty. You can see that in the crowd as the hands bob up and down and come to rest between the extremes of total yes and total no. And you can feel it when someone asks you a question: Do you buy it? Are you with us? Will you do it? Did you do it? Understand?

The audience example shows that fuzzy sets are real. Fuzzy sets live outside of math. They live in every net we throw around every chunk of the universe. To some degree we all belong to every grouping and belief system and fad and cult and trend. We all subscribe to some degree to every political party, every sexual orientation, every lifestyle, every side of the argument. We are all left, right, center, straight, gay, bi, cool, square, plain, for, against, and indifferent. We may not know which degree we are which but we know we are all to some degree. Movies and books and art help us see the many sides of our selves. We grow and learn as we age and that just means our degrees change, our *fits* or *fuzzy units* change as we move from thing to not-thing, from belief to disbelief, and maybe back again. Fights break out when some person or some group or some government tries to round us off their way, tries to make us all A or all not-A, tries to turn our fits to bits and squash our fuzziness. In this sense voting just asks for trouble.

The audience example also walks us closer to the geometry of a fuzzy set. First we translate the voting to symbols and then translate the symbols to the math picture. Suppose 25 persons sit in the audience and we ask them if they are married. Say 10 hands go up and 15 stay down. The answers are binary. The bit numbers 0 and 1 describe them. A 25-place list of bits describes the entire audience answer. If the person in the third seat raises her hand, a 1 goes in the third slot. Else a 0 goes in the third slot. The bit list

1 0 1 0 0 0 1 1 0 1 1 0 0 0 0 0 1 1 0 1 0 0 1 0 0

shows one way the 25 hands could have raised and not raised. There are more than 33 million (2^{25}) ways the hands could have raised or not raised. Each bit list defines one of the corners of the 25-dimensional hypercube with over 33 million corners. You can't picture cubes with more than three dimensions and eight corners and you don't need to to see the fuzzy structure. Bit lists or cube corners have no fuzzy structure. They are black and white, Aristotle's A OR not-A holds 100%, and they have zero fuzzy entropy.

Now consider a round of trap in the gun park. You yell "Pull!"

and the judge pushes a button and the trap box throws out a clay pigeon at 45 m.p.h. and you point and shoot a 12-gauge shotgun at the clay as it spins away from you in a high arc out over a field. You do this 25 times to use up the 25 shells in the box. You hit each clay or miss each clay or so the judge says. The judge shouts "Lost!" if he thinks you missed and then he writes a 0 on the bit-list scorecard. He says nothing and writes a 1 if he thinks you hit the clay. A good shot powders the clay. The judge always calls these as hits. A cheap shot nicks the clay and knocks off only a small piece of it. It breaks the clay only to some degree.

These are fuzzy hits and the judge can call them as hits or misses. There is little at stake and little honor in arguing for a nick hit while the next shooter waits to shoot. So few shooters do more than turn and look at the judge or shake their heads. Most just bite the bivalent bullet and let the judge round them off. When the round ends the judge calls the scores in hits out of 25 shots and each shooter gets to look at her 25-bit score list. Again your score list puts you at one of the over 33 million corners of a hypercube of 25 dimensions. And again there is no fuzzy entropy. The hit set is A and the miss set is not-A and the two sets do not overlap or underlap. But the clays broke only to some degree and not all or none. So there was fuzzy entropy or vagueness or "system disorder" in the round of trap even if there was none in the score.

Suppose the judge scored each shot as HIT or MISS or NICK and wrote a 1 or 0 or ½ on the score card. I have proposed trivalent scores to judges and shooters but no one wants them. They admit it is more accurate but it complicates the game and the stakes are so low that they prefer a simpler score to a fairer one. I also think most score totals would fall if judges wrote ½ instead of 1 for the many poor clay breaks. But suppose the judge scores nicks as NICKS and writes ½s on the score list. Now there are many more possible score lists and most of them contain some fuzzy entropy. Before, there were over 33 million (2^{25}) bit scores or cube corners. Now there are almost a trillion (3^{25}) possible score lists and all but 33 million or so are fit scores that include at least one NICK or midpoint score point. The fit scores lie *inside* the fuzzy hypercube of 25 dimensions with over 33 million corners. The new trillion or so points define a lattice within the fuzzy cube. And the score list of all nicks (½ ½ . . . ½) defines the midpoint of the cube and has fuzzy entropy 100%. All the other fit scores have fuzzy entropies less than 100% but more than 0%.

The fuzzy cube fills in if the judge allows more scores. Let's look at this with one shot of doubles. You yell "Pull!" and this time the trap throws out two clays at different angles and you shoot and swing and shoot. In the bit case there are four outcomes. You can miss both or (0 0). You can hit both or (1 1). Or you could hit the first clay and miss the second (1 0) or miss the first and hit the second (0 1). Two clays give a cube or square of 2 dimensions and 4 corners or a binary lattice (Figure 9.3).

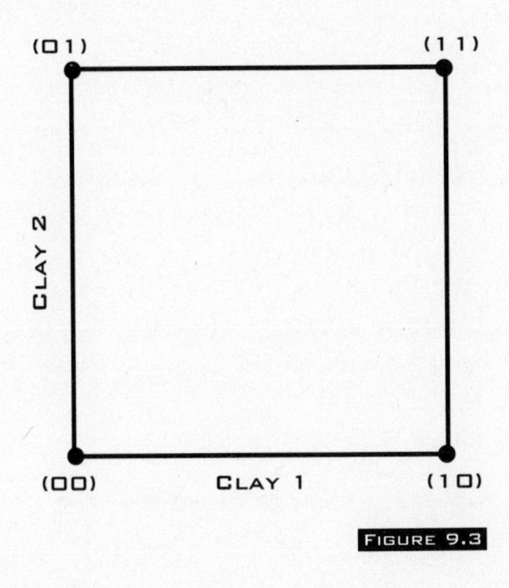

FIGURE 9.3

This bit cube is not a cube at all. It is just the 4 corners of a square, the grin of the binary cat. No points lie inside the square. This leaves out the midpoint and infinitely many other points.

Now suppose the judge scores the two shots as HIT, MISS, or PARTIAL—as 1, 0, or ½. This gives 9 (3^2) possible scores. Five of the 9 score lists are fit scores and the midpoint (½ ½ ½) is one of them. The score lattice fills in the cube a little more (Figure 9.4).

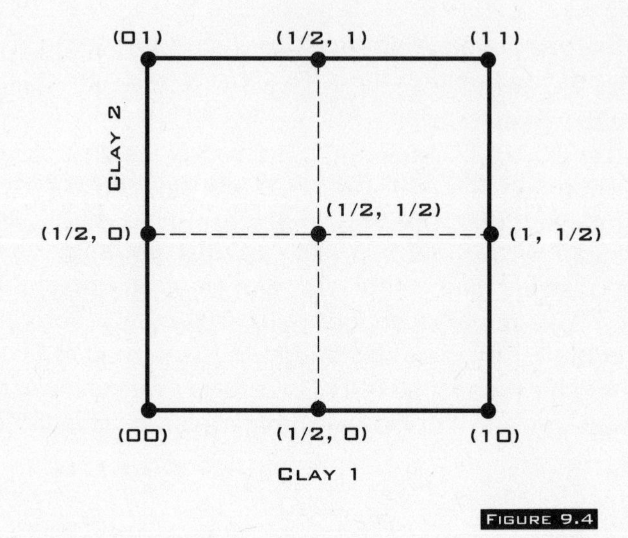

FIGURE 9.4

Now suppose the judge scores the two shots with 5 possible scores: MISS, NICK, PARTIAL, MOST, and HIT, or 0, ¼, ½, ¾, and 1. This gives $25(5^2)$ possible score lists and a denser fuzzy lattice (Figure 9.5).

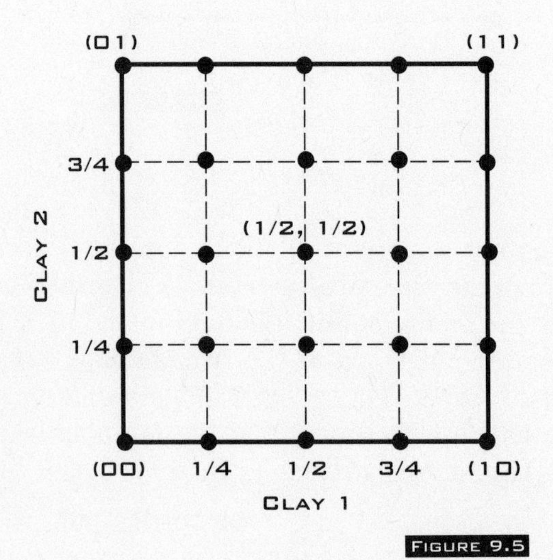

FIGURE 9.5

The four bit corners have 0% fuzziness. The midpoint has 100% fuzziness. The other 20 points have fuzzy entropy values between these two extremes.

The judge can allow more and more scores until he allows the whole continuum between 0 and 1. Then the cube fills up and becomes a solid cube. The cube still has only one midpoint and that still has 100% fuzzy entropy because there the yin-yang equation holds (A = not-A) 100% and there the Buddha's law of overlap A AND not-A holds 100%. The cube corners stay black and white and have 0% fuzzy entropy because there Aristotle's law of underlap A OR not-A holds 100%. All the other cube points are fuzzy or gray to some degree. Fuzzy entropy shows that *the closer to the midpoint, the fuzzier the set.* And the closer to a corner, the less fuzzy.

You can measure fuzzy entropy with two strings. Pick a point in the fuzzy cube and call it A. In the trap case take the score (⅔ ¼) where you almost powder the first clay but only nick the second. Find this score or fuzzy set A as a point in the cube (Figure 9.6).

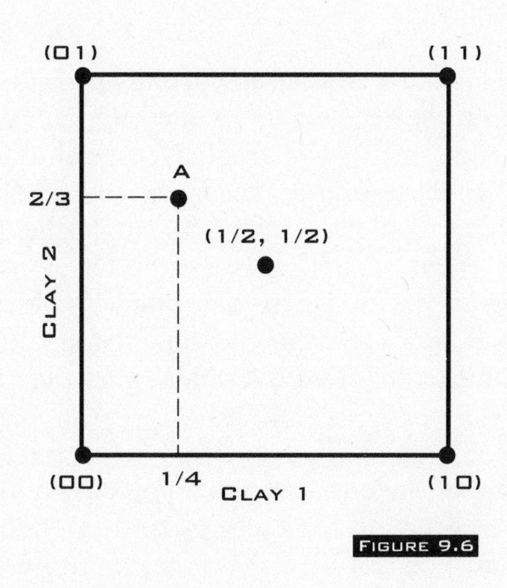

FIGURE 9.6

Now tie a red string from A to the *nearest* corner. This string keeps track of how close you are to the corner and how far you are from the midpoint. But all must be in balance. If A moves away from the nearest corner, it moves closer to the *farthest* corner and vice versa. So tie a blue string from A to the farthest corner (Figure 9.7).

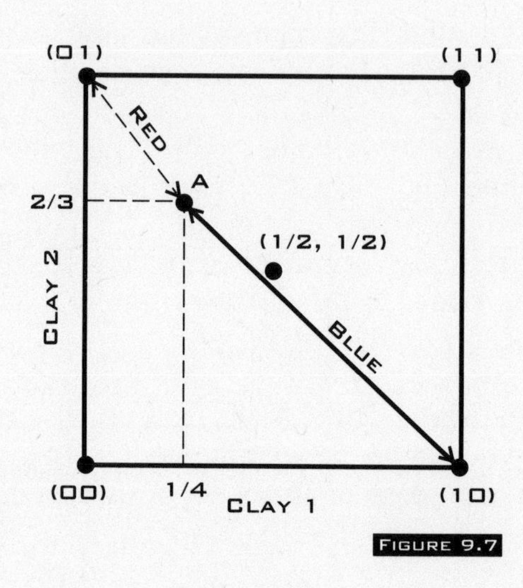

FIGURE 9.7

Then the percent measure of fuzzy entropy is just the red over the blue. Divide the red length by the blue length and you get the fuzzy entropy. You get a number between 0 and 1 that describes the vagueness of A. The bigger the number, the more the vagueness or fuzziness. In this example the fuzzy entropy equals 7/17 or about 41%. That is not too high since the shooter almost powdered the first clay and almost missed the second clay. And what holds for the two-dimensional cube holds for cubes of all dimensions. Red over blue, near over far, measures fuzziness.

Look at the extremes. Suppose you hit or miss all clays and so A lies at one of the corners. Then you cannot tie a red string to the nearest corner. The length is zero. Divide zero by any blue length and you get 0 or 0% fuzzy entropy. At the other extreme you half break each clay and A lies at the midpoint. Then every corner is equally close and equally far. So the red equals the blue. So the ratio equals 1 or 100% fuzzy entropy.

You can see fuzzy entropy in any room. Stand in a room or office or hall and see the place as a cube of three dimensions.

See your fist as a point in the room. Find the corner closest to your fist. Look away from this corner to the other side of the room and you will find the corner farthest away from your fist. A long diagonal line connects the two corners. See or tie a red string from your fist to the closest corner and a blue string from your fist to the farthest corner. Divide the red string length by the blue string length and you have found the fuzziness of your fist. If the red string is 5 feet long and the blue string is 15 feet long, the fuzzy entropy is ⅓ or 33%.

Next comes the fuzzy entropy theorem. I proved it one afternoon in January 1986 as I tried to make up a three-hour lecture for a fuzzy course I had to teach that night at UC San Diego. I sat in a small cube of a room and stared at the white board on the wall. I felt there was no good book on fuzzy sets that you could teach out of. I needed easy homework problems to work in class and to assign for next week and had to come up with my own. I had just found the red-blue measure of fuzziness the month before and wanted to look at it some more with homework problems. The students had liked my use of fits and bits the week before and so I wrote out some toy problems using short fuzzy sets or short fit lists.

Soon I got the same answers on different problems. I wrote out some more problems and answers and then wrote the equation on the white board and its picture next to it. I got excited and rushed the problems and wrote so fast and so badly that later I could not read it. Then I stated the fuzzy entropy theorem. I guessed at it. I wrote it down in red on the white board and looked at it. It had to be true. The symmetry was too perfect. Then I wrote out the proof and confirmed my intuition. It was true, is true, will always be true. The Buddha over Aristotle.

The fuzzy entropy theorem says the Buddha is red and Aristotle is blue.* The red string over the blue string equals the

*The Fuzzy Entropy Theorem gives the entropy $E(A)$ of fuzzy set A as the ratio of the counted overlap or intersection $A \cap A^c$ to the counted underlap or union $A \cup A^c$:

$$E(A) = \frac{c(A \cap A^c)}{c(A \cup A^c)}$$

A^c denotes the complement of A or the set not-A. With nonfuzzy sets the overlap is empty and so the numerator equals zero. With fuzzy sets there is overlap and the numerator is always greater than zero. So fuzziness begins where Western logic ends.

Buddha over Aristotle, A AND not-A over A OR not-A, overlap over underlap.

Think of it this way. If you have A, you have not-A. So you also have A AND not-A. And you also have A OR not-A. Once you have one set or point you have all four and all four are equally close to their nearest corner. The theorem uses the length or count of these sets. (You count the set ($\frac{2}{3}$ $\frac{1}{4}$) by adding the fits: $\frac{2}{3} + \frac{1}{4} = \frac{11}{12}$. You add up the fits just as you add up the 1s and 0s in a trap round to get your score.) The red string equals the count of the Buddha term A AND not-A. The blue string equals the count of the Aristotle term A OR not-A. In the cube you can see these lengths are equal (Figure 9.8).

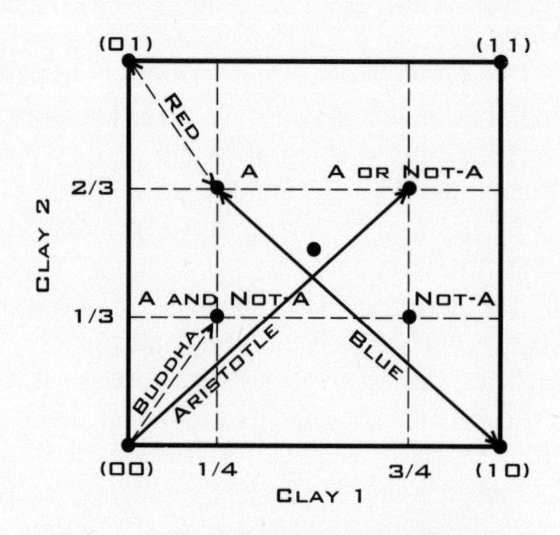

FUZZY ENTROPY THEOREM

FIGURE 9.8

The two dashed lines have the same length and so do the two solid lines. If we choose A closer to the midpoint, then the other three points move closer to the midpoint as well. So the fuzzier the set, the more the Buddha's A AND not-A resembles Aristotle's A OR not-A. At the midpoint all four points are equal. A equals not-A and the yin-yang equation holds 100% and the fuzzy entropy is maximal. At the midpoint you cannot tell black from white or white from black. The midpoint is the black hole of set theory. It is the *gray* hole of set theory.

Fuzzy entropy holds another surprise: It equals how much the Buddha contains Aristotle. The whole in the part. Underlap stuck in overlap. A OR not-A stuck in A AND not-A. *Subsethood.* In Chapter 3, The Whole in the Part, I argued that subsethood or degree of containment is the deepest and strangest idea in fuzzy logic and it explains probability or "randomness" as the whole stuck in the part. Everything reduces to subsethood. Fuzzy entropy reduces to subsethood too. The fuzziness of set A equals how much the overlap A AND not-A contains the underlap A or not-A.* This is odd since the overlap is part of the underlap. The underlap is part of the overlap only to some degree. The whole in the part is the essence of fuzzy logic.

This whole section on fuzzy entropy has been about A AND not-A. This whole book is about A AND not-A. Shades of gray come from the overlap of A AND not-A. Common sense and wisdom and confusion also come from it. Fuzzy entropy takes these ideas into equations where they have never been before. It measures the shadow that a fuzzy set casts on the math wall.

So far I have talked only about *my* math work on fuzzy sets. But I was not here first. I did not "invent" or first discover fuzzy sets. Two other men did that. Max Black first had the idea. Later Lotfi Zadeh worked out the formal math system of fuzzy sets and gave them their name and fought for them for years until the Japanese first put them into machines and made fuzzy sets rich and famous.

MAX BLACK: VAGUE SETS

In 1937 quantum philosopher Max Black published a paper called "Vagueness: An Exercise in Logical Analysis" in the

*This theorem gives fuzzy entropy E(A) as a degree of subsethood:

$$E(A) = S(A \cup A^c, A \cap A^c) = \text{Degree}(A \cup A^c \subset A \cap A^c).$$

The subsethood term $S(A \cup A^c, A \cap A^c)$ measures the degree to which the underlap $A \cup A^c$ is a subset or part of the overlap $A \cap A^c$. The whole $A \cup A^c$ always contains the part $A \cap A^c$. So $S(A \cap A^c, A \cup A^c) = 1$. But the part contains the whole only to some degree. So $S(A \cup A^c, A \cap A^c) \leq 1$.

journal *Philosophy of Science*. Black's paper defined the first simple fuzzy set with what we now call a membership curve. Points on the curve give the fit values of elements in the fuzzy set A or the set not-A (Figure 9.9).

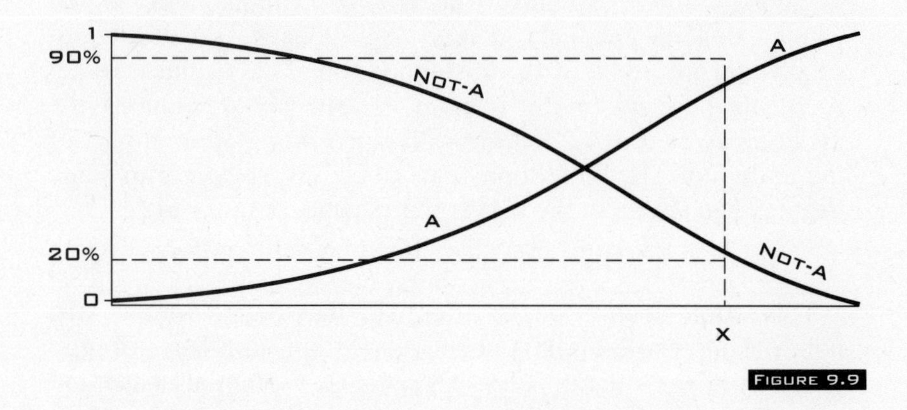

FIGURE 9.9

The A and not-A curves* are the meat of fuzzy set theory. Black saw that everything is to some degree A and to some degree not-A. Everything is red and not-red, large and not-large, smooth and not-smooth. The degrees may be so small or large that we take them as none or all, as 0 or 1. So Black drew these as curves that approach the extremes of 0 and 1 but need not reach them. The curves are not abrupt steps from 0 to 1 or from 1 to 0.

The curves show another point of pure fuzz. They show that not-A is the *inverse* of A and vice versa. If you add the A curve to the not-A curve you get a flat line at the 100% value. So not-(not-A) is A. If the A curve touches 1, the not-A curve must

---

*You can view these curves as the limiting case of points in larger and larger fuzzy cubes. If a cube has only two dimensions, then the "curve" looks like two spikes.

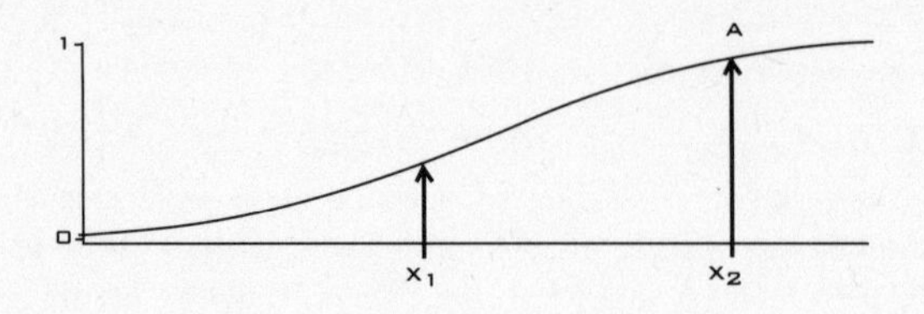

touch 0 and vice versa. The curves move away from the extremes as they get fuzzier or vaguer. They cross at the midpoint fit value of ½. As Black put it, "Thus the very precise symbol [set] would have a consistency [fuzzy] curve made up of a straight line almost parallel to the horizontal axis, and at a great distance from it, followed by a steep drop to another line almost parallel to the horizontal axis and very close to it." Note that the two curves cross at the midpoint value ½ where A equals not-A.

If you take, say, 25 samples from each curve, then each list of samples is the same as a fuzzy trap score and defines a point in a fuzzy cube of 25 dimensions. Again as you increase the number of samples you increase the cube's dimensions. Black drew curves but worked with samples or fit lists: "In practice the number of terms in A will usually be very much greater than 10 (e.g., there are said to be something like 700 distinguishable *shades of gray*) and the consistency [fuzzy set] curve will approximate to a smooth curve having a continuous gradient."

Max Black used the term "vague" because Charles Pierce and Bertrand Russell and other logicians had used it to describe what we now call "fuzzy." Consider this fuzzy quote, which foreshadows this whole book, from systems theorist Jan Christiaan Smuts in his 1926 book *Holism and Evolution*:

If a cube has 10 dimensions, then each point in it is a fuzzy set with ten fit values. So the "curve" of each fuzzy set looks like ten spikes.

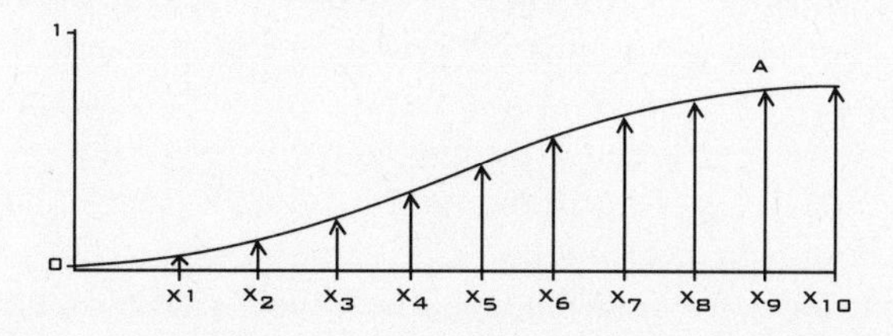

And so on. Add more dimensions to a cube, let the cubes grow larger, and you get more spikes. In the limiting case the spikes are so dense they almost equal a real curve. (The spikes are "dense" in the curve as the rational numbers are dense in the number line. Near any number you can always find a rational number, a ratio of whole numbers, that is as close to the number as you wish.) In this sense fuzzy cubes *digitize* fuzzy curves or vague curves. In some sense we only work with a finite set of numbers in the real world and in computers. So we always work with some digitized fuzzy set, some point or fit list in some fuzzy cube.

The science of the nineteenth century was like its philosophy, its morals and its civilization in general, distinguished by a certain hardness, primness and precise limitation and demarcation of ideas. *Vagueness*, indefinite and blurred outlines, anything savoring of mysticism, was abhorrent to that great age of limited exactitude. The rigid categories of physics were applied to the indefinite and hazy phenomena of life and mind. Concepts were in logic as well as in science narrowed down to their most luminous points, and the rest of their contents treated as nonexistent. Situations were not seen as a whole of clear and *vague* elements alike, but were analyzed merely into their clear, outstanding, luminous points. A "cause," for instance, was not taken as a whole situation which at a certain stage insensibly passes into another situation, called the effect. No, the most outstanding feature in the first situation was isolated and abstracted and treated as the cause of the most outstanding and striking feature of the next situation, which was called the effect. Everything between this cause and this effect was blotted out, and the two sharp ideas or rather situations of cause and effect were made to confront each other in every case of causation like two opposing forces. This logic precision made it impossible to understand how the one passed into the other in actual causation.

There is no way out of this impasse but to retrace our steps and see that these concepts are partial and misleading abstractions. We have to return to the fluidity and plasticity of nature and experience to find the concepts of reality. When we do this we find that round every luminous point in experience there is a gradual shading off into haziness and obscurity. A "concept" is not merely its clear luminous center, but embraces a surrounding sphere of meaning or influence of smaller or larger dimensions, in which the luminosity tails off and grows fainter until it disappears. Similarly a "thing" is not merely that which presents itself as such in clearest definite outline, but this central area is surrounded by a zone of intuitions and influences which shades off into the region of the indefinite [A AND not-A]. The hard and abrupt contours of our ordinary conceptual system do not apply to reality.

The world is thus in [bivalent] abstraction constituted of entities that are discontinuous, with nothing between them to bridge the impassable gulfs, little or great, that separate them from each other. The world becomes to us a mere collection of *disjecta membra*, drained of all union or mutual relations, dead, barren, inactive, unintelligible. And in order once more to bring relations into this scrap heap of disconnected entities, the mind has to

> conjure up spirits, influences, forces and what not from the vast deep of its own imagination. And all this is due to the initial mistake of enclosing things or ideas or persons [or sets] in hard contours which are purely artificial and are not in accordance with the natural shading-off continuities which are or should be well known to science and philosophy alike.

The Smuts quote shows that the fuzzy or vague world view was alive and well in the early twentieth century. Max Black did not claim to invent the field. The scholars of the day knew that vagueness meant shades of gray and A AND not-A. Smuts also saw, as Max Black did not see and later Lotfi Zadeh did see, that vagueness applied to systems. Smuts helped found the holistic movement that led to modern systems science.

So what did Max Black claim to have done in his article? He put vagueness in symbols at the level of sets or systems. He felt he had met Henry James's challenge, which he cites as an opening quote in his article: "The notation, however, is what we lack, and the verdict of the mere feeling is liable to fluctuate." The notation. The formal theory. The math behind the idea. Fuzzy set curves. Black extended multivalued logic to sets and showed that these "fuzzy" sets matched our ideas of ideas.

In the end Black's theory did not sell. It died in silence in an elite journal that only a few men and women read. Dust storms and price deflation swept the American Midwest and the world prepared for its worst war and philosophers played with the new toy of black-and-white symbolic logic. Logical positivism peaked in the 1930s and 1940s and Black's article went up against the whole trend and lost. And it lost in silence. Black never pushed it again and the world of multivalued logic, which Black had learned from and tried to excite, paid no attention. The world of math never heard of Black's vague sets. Living philosophers have little power in the world of science.

Max Black died in August of 1989. He died the year that the fuzzy or vague logic he had studied first played on TV commercials for smart washing machines on Japanese TV. He died as his theory of vague sets had died—in silence. The world of philosophy did not credit his work on vagueness. Nor did the fuzzy world. Most fuzzy engineers still have not heard of Max Black or any of the other men who first worked out the logic of "fuzzy" logic.

It took a man of political and scientific skill to sell vagueness to science. Such individuals are rare. You find them in each branch of science and they tend to be the names you know in science or in any other field. No idea jumps from void to paper to common knowledge. In each case one man or woman adds a new twist and finds a new way to sell the idea. In this case it began with a name change from "vague" to "fuzzy" and an Iranian immigrant professor called Lotfi A. Zadeh. Max Black was born in 1909 in the city of Baku on the Caspian Sea. In 1921 Lotfi Zadeh was also born in Baku and that was the second thing the two men had in common.

LOTFI ZADEH: FUZZY SETS

In 1965 Professor Lotfi Zadeh published the paper "Fuzzy Sets" in the journal *Information and Control*. That year I was five years old. Zadeh was an associate editor of the journal and that helped get the paper published. You don't see many papers on set theory in engineering journals. Zadeh was a tenured professor in UC Berkeley's department of electrical engineering and computer science, the top or nearly the top school in engineering. More than that, Zadeh was chair of the department. That too helped get the paper published.

Zadeh wrote only fuzzy papers after the first one. In 1968 he left the engineering side of the department and went to the "softer" side of computer science. He was at the height of his academic power when he published "Fuzzy Sets." No one paid much attention to it. But Zadeh was a good speaker with a large following. He had co-authored in 1963 the first book on linear systems theory. Every engineering school used it as a text. So when Zadeh came to give a talk in a department or at a conference he had an audience. When he wrote more papers on fuzzy sets more people read them.

In effect as Zadeh moved into computer science he cashed in his engineering chips and spent them on his new fuzzy dream. He had tenure and all the status you could have in engineering. Now he set out to build a field. He knew it would take years and he had the years. He shifted more from math and number crunching to philosophy and language. In an age when scientists in every field of science threw more math at their problems, Zadeh threw less. At first he lost more friends than he gained. Young engineers who had taught with his text or who had been taught with it now could write off the big man as soft and fuzzy-headed. Critics demanded that he show that fuzziness was not probability in disguise. Zadeh did not bother with that fight. When pushed at a conference talk he would point to how we use language when we reason or when we build or model a system. Some heads would nod and some would shake.

Consider this quote from Professor Rudolf Kalman in 1972 at the Man and Computer Conference in Bordeaux, France. Zadeh talked on fuzzy sets and systems and then a moderator asked Kalman to respond. Kalman, who designed the "Kalman filter," used more than any other system in navigation, guidance, estimation, and in hundreds of other computer tasks, had this to say:

> No doubt Professor Zadeh's enthusiasm for fuzziness has been reinforced by the prevailing political climate in the U.S.—one of unprecedented permissiveness. "Fuzzification" is a kind of scientific permissiveness; it tends to result in socially appealing slogans unaccompanied by the discipline of hard scientific work and patient observation.

Talk like that from Kalman would have ruined many a lesser career. It bounced off Zadeh. Zadeh always says that he has a thick skin and he does. Many times I have seen him attacked and insulted in public. He just laughs.

The Kalman quote runs deeper than it looks. The two men have a history. Kalman was a graduate student at Columbia University in the late 1950s when Zadeh taught there. The two worked in the same circles and were friends but were never too close. Both worked in systems theory. Zadeh's work on "frozen poles" and "shaping filters" helped lay the foundation for the Kalman filter. But Zadeh missed the prize and Kalman found it. That's why we call it the Kalman filter and not the Zadeh filter.

Pound for pound and bit for bit the Kalman filter belongs with the light bulb and the radio and the microprocessor in its engineering benefits for mankind. In many ways it goes beyond these since it is just a set of equations. Like all math the Kalman filter is eternal. Alfred Nobel did not give engineers a chance at a Nobel Prize. If he had, Kalman would have won hands down and no one would grouse about it. Yet most people have never heard of the Kalman filter. It guides airplanes and space probes and cruise missiles and tracks satellites and economic trends and changes in your bloodstream. It is an "optimal" estimator. It gives the "best" guess where the plane went when it flew behind a cloud. Didn't you ever wonder how a missile finds its target or how the astronauts find their way home? The Kalman filter shows them how. It stands to much of engineering as Darwin's theory of natural selection stands to evolutionary biology. Thousands of engineers have published small tweeks and variants of the Kalman filter. And they and tens of thousands of other engineers have dreamed what it would be like if they had been the first to find and publish the optimal filter. All the glory and prestige and money—even though Kalman filed no patent.

I think at some time in the early 1960s Zadeh had that thought too. I know him well and asked him one day about it. He just waved it off and said the Kalman filter was "too Gaussian." That means it depends too much on a bell curve. And it does. But the bell curve fits many cases and if you had to pick a curve to work with there are many math reasons why the bell curve is the best curve you could pick.

Zadeh is one of the smartest and shrewdest men I know. From his high Berkeley perch he would have seen early on in the 1960s that Kalman had walked off with the one good plum in his field. Most of Kalman's followers did not see this until the 1970s or 1980s. And in the 1950s Zadeh had come close to the Kalman filter with his early work that helped make him department chair and helped make his text a classic in systems science. I think that is why he jumped into fuzzy sets and why he called "fuzzy" what Bertrand Russell and Jan Lukasiewicz and Max Black and others had called "vague" or "multivalued." Start fresh in a new field. Turn and challenge the math of systems theory. Better to rule in hell than serve in heaven.

I repeat that this is speculation and that Lotfi and I are good friends and have been for years. But for me it fits the facts. I have asked Lotfi about it and he has never given me a full answer. He has told me how he came to America and how he came to the idea of fuzzy sets and this I will tell you below. But as Lotfi says he remains true to his Persian roots. You cannot pin him down just as you cannot get him mad. He has the political genius of an aikido master. He stays cool and smooths feathers and deflects blows and above all stays afloat. Many times he has helped me and other friends out of tough spots. He leads by character and never by force and plays no dominance games. He charms both women and men. Zadeh's favorite quote: "Friends come and go but enemies accumulate." At a workshop or restaurant he invites everyone to join him in a talk or dinner. He is a free spirit in every sense of the term.

Zadeh lives in a world of technical talks and evening parties. He thrives on this and flies round the world to keep it up. Above all else he loves the company of people young and old.

I remember the summer of 1987 after I had just sent Zadeh a copy of my Ph.D. dissertation on fuzzy systems. I lived then with my new wife and our baby daughter. One Friday night Zadeh was in town and called my house looking for me. I was away giving a seminar on neural networks and fuzzy logic at Wright Patterson air force base in Ohio. My wife answered the phone and told Lotfi I was out and that she was going to a party at a girl friend's house. They had never met but Zadeh said he had nothing to do and persuaded her to pick him up and take him with her. So they went to the party. Zadeh had almost 40 years on everyone there but did not let that faze him or slow him down. The next day I returned and that night he called me to see how things were. I said things were fine and hoped I had that kind of energy when I turn 67.

When Zadeh was born in 1921 in Baku, it was the capital of Soviet Azerbaijan. Zadeh was an Iranian citizen but his first language was Russian. His father was a businessman and a correspondent for an Iranian newspaper, and Zadeh lived in Iran from age 10 through age 23 and went to a Presbyterian school. The Zadehs were well-to-do, and young Lotfi always had his own car and his own servant. In 1942 he graduated with a bachelor's degree in electrical engineering from the University of Tehran. He used war connections with the U.S. Persian Gulf Command to get one of the 100 U.S. immigration slots.

In 1944 he arrived in the U.S. and went to MIT, where in 1946 he got a master's degree in electrical engineering. At that time his parents moved from Iran to New York City. Zadeh left MIT and joined his parents in New York and enrolled in Columbia

University. In 1951 he got his Ph.D. in electrical engineering, and joined Columbia's faculty. He stayed there until he moved to UC Berkeley in 1959. In 1963 he became chairman of UC Berkeley's electrical engineering department. That's as high as you get in engineering. Twenty years before he had given orders to servants in the Middle East. Now he hired, reviewed, promoted, and fired some of the top engineering faculty in the world.

Zadeh's iconoclasm got him into multivalued logic. He saw his colleagues rush deeper into advanced math to describe physical and engineering problems. Zadeh is the kind of man who always considers the contrary and tends to argue for it. He told me that he began thinking about multivalued logic when he was at Columbia University in the early 1950s. At that time he directed his graduate student Oscar Lowenschuss in writing one of the first engineering dissertations on how to build multivalued-logic electrical circuits.

Zadeh's next move in the fuzzy direction came in 1956, when Princeton's Institute for Advanced Studies invited him to visit for a year. His friend Herbert Robbins, the founder of an important branch of statistics called stochastic approximation, invited him. The institute housed a pantheon of great mathematical thinkers that included Albert Einstein and Kurt Gödel. There an electrical engineer was, as Zadeh put it, the odd man out. An engineer still is the odd man out in the midst of physicists and mathematicians. At seminars and parties Zadeh said he was an applied mathematician.

At the Advanced Institute Zadeh met Stephen Kleene, who was at Princeton, the leading multivalued logician in the United States. Kleene took the young Persian thinker under his wing. They wrote no papers together but they influenced each other. Zadeh learned formal logic and the math of multivalued logic. Kleene had come up with much of that math. Zadeh taught Kleene the principles of electrical engineering and information theory.

Zadeh went back to Columbia and on to Berkeley, sold on multivalued logic. He had seen the black and the white reduced to a special case of gray. Kleene had helped him weave this into his world view. And right away Zadeh set out to apply it to systems theory, the hot topic of that day.

Here Zadeh made his deepest contribution. It was an idea and an approach and not the later math of fuzzy sets. His colleagues used ever more detailed math and peered closer and closer at complex systems in engineering, economics, weather forecasting, and biology. Zadeh stepped back. He saw that as the system got more complex precise statements had less meaning. He later called this the *principle of incompatibility*: Precision up, relevance down. In 1962 Zadeh published this change of heart in the paper "From Circuit Theory to System Theory" in the *Proceedings of the IRE*, the leading engineering journal of its day. Here Zadeh first suggested the term "fuzzy" for multivalence and came close to calling the whole mess "cloudy" instead of "fuzzy," in which case this book would bear a different title:

> There is a fairly wide gap between what might be regarded as "animate" systems theorists and "inanimate" systems theorists at the present time, and it is not at all certain that this gap will be narrowed, much less closed, in the near future. There are some who feel this gap reflects the fundamental inadequacy of the conventional mathematics—the mathematics of precisely-defined points, functions, sets, probability measures, etc.—for coping with the analysis of biological systems, and that to deal effectively with such systems, which are generally orders of magnitude more complex than man-made systems, we need a radically different kind of mathematics, the mathematics of fuzzy or cloudy quantities which are not describable in terms of probability distributions. Indeed, the need for such mathematics is becoming increasingly apparent even in the realm of inanimate systems, for in most practical cases the *a priori* data as well as the criteria by which the performance of a man-made system is judged are far from being precisely specified or having accurately known probability distributions.

Zadeh saw his colleagues caught up in a mathematical arms race. They attacked ever harder problems and competed by throwing more and more math at them. He wanted to sidestep that arms race and work with big common-sense chunks. He was also versed in multivalued logic and almost no one else in engineering was. So he mixed the two and called it fuzzy set theory. He now had a new unit to work with. He had the fuzzy set as a chunk of common sense. He could have used some other theory of uncertainty to state the same thing. But multivalued theory was at hand. The fuzzy set was born.

The fuzzy set that made the case was the set of tall men. Zadeh tied the concept TALL to a curve of fit values (Figure 9.10). Zadeh called this fuzzy set a *membership curve*. It acts just like a vague list or curve of Max Black. For each height it gives the degree or fit or membership in the set of tall men. Tallness is a smooth function of height. Every man is tall to some degree. Every man is not tall to some degree. So NOT TALL looks like the inverse of the TALL curve (Figure 9.11). If you draw the TALL and NOT-TALL curves on the same graph, you can see that, as with Black's vague lists, they intersect at the midpoint fit value of ½ where A equals not-A.

Zadeh went on to show how fuzzy sets give a linguistic calculus. You can put all the modifiers you want in front of TALL. VERY TALL shrinks the TALL curve. MORE OR LESS TALL raises the TALL CURVE (Figure 9.12).

You can do the same thing with the set of old houses or very old houses. Zadeh saw that if curve A never rises above curve B, as VERY TALL never rises above TALL since it takes more height to be very tall than just tall, then A is a 100% subset of B. He stopped short here and missed the theory of fuzzy containment. A can still be a subset of B to large degree if the A curve rises above the B curve in places.

Part of the power of the fuzzy curves came from how silly it made nonfuzzy sets look. They were step functions or hard lines drawn between A and not-A. The old view of TALL was bivalent. There were men in the set of tall men and men in the set of not-tall men and no men in both and no men who were not in one. You jump from not-tall to tall at some height, such as six feet. You draw a hard line at six feet between tall and not-tall (Figure 9.13).

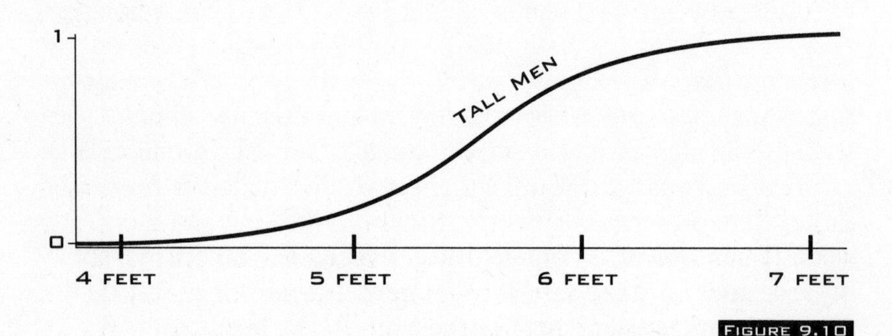

FIGURE 9.10

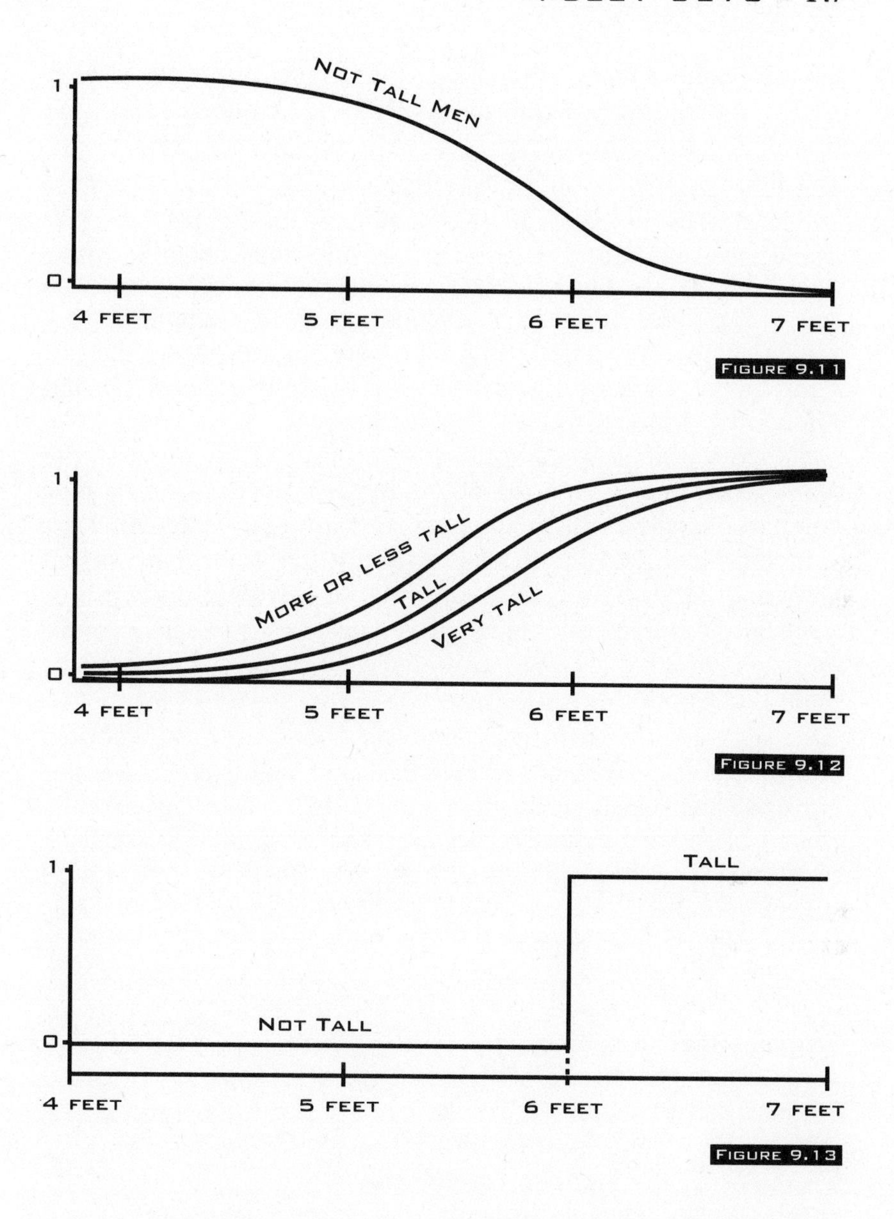

FIGURE 9.11

FIGURE 9.12

FIGURE 9.13

But tallness, like most properties of the world, is a matter of degree. It grows with height. A curve shows this smooth change. A hard line can never show it. That again is the advance of fuzzy sets. It ties words to curves (or to fit vectors in a fuzzy cube).

The math of fuzzy sets was not new. It used the same algebra that Jan Lukasiewicz had worked out half a century before in his multivalued logic. And it used the vague list or set of Max Black. Zadeh did not call his new sets multivalued sets. That would have been the natural name in light of Lukasiewicz's work and the fact that a whole field went by the name multivalued logic. Or he could have called his sets vague sets after Russell and Black. But he chose "fuzzy." I have asked him why many times. He said he chose "fuzzy" because of its ties to common sense. He felt it captured the vague notion of a chunk that he was after. At the same time his first and future papers did not refer to the work of Lukasiewicz or Russell or Black (he did refer to his friend Kleene). That suggests there may have been some marketing at work in the choice of name. In any case the name stuck and Zadeh and others began to build a whole new computer science or language science with their fuzzy chunks.

In the next decade Zadeh drifted further from his roots in math and hard science. He went into the new field of computer science but still kept his old goal of a new systems science built on fuzzy sets. By 1972 he had stated the principle of incompatibility this way:

> As the complexity of a system increases, our ability to make precise and significant statements about its behavior diminishes until a threshold is reached beyond which precision and significance (or relevance) become almost mutually exclusive characteristics. . . . A corollary principle may be stated succinctly as, "The closer one looks at a real-world problem, the fuzzier becomes its solution."

Precision increases fuzziness.

As Zadeh spread the fuzzy word the criticisms grew. These fell into three categories. The first was a call for a novel application. What can you do with fuzzy sets? Show us a working system or simulation. Do something with it.

New ideas take time to apply. For years Zadeh and his few fuzzy followers could show no applications. And Zadeh never did. He published "Fuzzy Sets" in 1965 as a math theorist and he stayed a theorist. He just switched the theory from math and systems science to linguistics and cognitive psychology. In the 1970s fuzzy applications appeared but these were often toy computer simulations of simple math ideas. The one exception was the systems work of Ebrahim Mamdani in England. That led to fuzzy systems as we know them now and as we will look at them in the next chapter. In the 1980s the Japanese pursued these systems for control. By 1990 the Japanese had over 100 real fuzzy control applications and products.

The second criticism came from the probability school. Zadeh used numbers between 0 and 1 to describe vagueness or matters of degree. Probabilists felt they did the same thing. A fight was inevitable. Just as with the first calls for applications, the probability critics overwhelmed Zadeh. He was one and they were thousands with hundreds of years of math and culture behind them. Most of these critics just dismissed fuzziness as randomness in disguise. They felt Zadeh had not made his case that fuzziness was new. And Zadeh did not really try. He focused on the expressive power of fuzzy sets and how well it fit with our words. That went over like telling a joke on the witness stand. Zadeh took a pounding from the probability school. That was still going on when I met him in 1984 at a conference on artificial intelligence. Zadeh sat on a panel and just grinned as a probabilist pounded him from the podium. In Chapter 3, The Whole in the Part, I discussed how this drew me into the fuzzy fight and how I soon saw that you could turn the tables on the probabilists and use fuzziness (or subsethood) to explain probability: probability as the whole in the part. But that came almost a quarter century after Zadeh published his first fuzzy paper.

There was one probability criticism that shook Zadeh. I think this was why he stayed out of probability fights. He and I have often talked about it but he does not write about it or tend to discuss it in public talks. He said it showed up as early as 1966 in the Soviet Union. The criticism says you can *view* a fuzzy set as a random set. Most probabilists have never heard of random sets. But in math there are many obscure niches and this was one. These critics did not dismiss Zadeh's ideas out of hand. They thought the ideas were strong. They just thought they were not his ideas.

Look at the first curve for TALL. You can view each number in the curve as the degree that a man of a given height is tall. The height values index the fit values. Tall *given* height. So why not view this as a probability conditioned on height? Pick a height, say six feet. Then the man is tall or not. On the fuzzy view he is tall, say, to degree 80% and so he is not tall to degree 20%. But you can also *say* that at six feet the man is tall with probability 80% and not tall with probability 20%. No doubt there are cases where that makes sense. In the next chapter I mention how I have used this probability view when I wanted to prove a probability theorem about a fuzzy system. It's just a complicated way of looking at things. For most of us when we say that a man is tall we don't mean it in the sense of a bingo game. His tallness is deterministic. It is a property of his being. It is a matter of degree.

In the first chapter I used the example of parking your car in parking spaces. It applies here too. Say you park on two spaces. You overlap two spaces, 80% on the first space and 20% on the second space. You can say that you parked on the first space with 80% probability and on the second with 20% probability. You can say that because language is sloppy and, most of all, because 80% and 20% add up to 100%. The probability words are consistent with the probability math. But they do not match fact. We do not mean that you parked completely within the first space and you did this with 80% probability and that you parked completely in the second space with 20% probability. We mean you parked in one space or the other and you don't know which. You just know there is an 80% chance you parked in the first space and 20% in the second space. Then the two spaces don't even have to be next to each other. That state of affairs will never get you a ticket for taking up two spaces. We mean overhang. Crossing the boundary. Degrees. Fits.

In the end Zadeh ignored this probability criticism perhaps because so few people made it. He is not the kind of man who has to beat every criticism to death. "Just water on a duck's back," he likes to say. He wrote his papers and gave his talks to show the nice fit of fuzzy sets with our fuzzy words. This did not answer his critics and he did not care. It got the fuzzy idea out. It made hundreds and then thousands of new followers. By the 1970s he had a philosophical cult and he was its guru.

The third criticism was the biggest one. It was the naked wrath of bivalence. Only a handful of thinkers read the vague work of Black or Russell. Thousands read the fuzzy work of Zadeh. His Berkeley position gave him a pulpit to spread the fuzzy word. Zadeh had his own way of telling the fuzzy story. He focused on how we use language and downplayed the math. But the math of multivalence was still there. Sooner or later his audience saw what he was saying: A AND NOT-A is okay.

For Zadeh to be right or even plausible, Aristotle had to be wrong. That meant things did not have to be black and white. Our bivalent truths were just conditioned reflexes. Three thousand years of Western history were just one of many possible paths through space-time. It might all be wrong or, worse, a limited cultural whim: science as precision, math and logic as lines between black and white. That was a lot to accept in one lecture or from one paper. Zadeh seemed to say that either he was right or most of science and Western culture was right. One or the other. The choice seemed clear.

I think he was right. That was why I became a fuzzy theorist and tried to flesh out the foundations of fuzzy sets and systems. I became friends with Zadeh in 1984 when I was just twenty-four. I sent him every new paper and math idea I had and we talked on the phone and met at conferences and workshops. I was one of the faithful. I saw liberation in fuzzy logic. I did not care for its cult side of internal Zadeh worship and anti-probability dogma and for many of the followers and their work. I learned a great deal from fuzzy decision theorist Ron Yager at Iona College and fuzzy information theorist George Klir at SUNY Binghamton. But most followers just followed and had no new ideas and did not challenge the old ones. They had not made their case. Too often their papers just preached to the fuzzy choir. They did not try to persuade the unpersuaded and publish in nonfuzzy journals. They were loyal to the cause and Zadeh always rewarded them with his friendship and help in publication and letters of recommendation. But they were not of Zadeh's caliber. They did not have his track record and standing and political skill and deep intuition. And what bothered me most was that they had joined a cult and just assumed that the cult's god existed. Every cult has a god and you join the cult when you believe in the god.

The probability group is a cult too, and a big one, and has its god of maximum probability. It is so big it counts as a math religion. The fuzzy cult was small and its god was fuzz. I felt you had to go through a "furnace of doubt," as Dostoyevsky called it, before you found real faith in your god. How can you back fuzzy logic when you know the probabilists laugh at it as confused or as probability in disguise? Can't you hear the bivalent philosophers at the door? Who is right? In the end you have to act on your own judgment and live with it. I spent years in doubt. First I doubted the bivalent school when I was 20 and still in school. When I was 23 I jumped into the fuzzy camp and the doubt got worse. When I was 25 I found the fuzzy cube and the early theorems on fuzzy entropy and subsethood and found peace. I had crawled out of the furnace and dug a place to rest with my own hands. I still thought Zadeh was right but now for different reasons. The world paid no attention. It still rolled down its bivalent groove.

The bivalent criticisms were of two kinds. The first kind says that bivalence works. Binary logic runs computers and has served us well for thousands of years. It may come with some costs but it is simple and it works. The second kind of criticism is a scream of outrage. It has no logical or factual content but it holds sway. It is the collective foot stamp of modern science as it denies A AND NOT-A and insists on A OR NOT-A.

Harvard philosopher Willard Van Orman Quine champions the first claim. He thinks bivalent is best. It makes sense to him and has guided science well and has the virtue of simplicity. Quine knows that this makes him draw a hard line at some height to split TALL from NOT-TALL. He does not know where to draw the line and admits it. But he thinks you can live with a floating boundary. This is how he put it in his famous 1981 article "What Price Bivalence" in the *Journal of Philosophy*:

> If the term "table" is to be reconciled with bivalence, we must posit an exact demarcation, exact to the last molecule, even though we cannot specify it. We must hold that there are physical objects, coincident except for one molecule, such that one is a table and the other is not.
>
> One might then despair of bivalence and proceed disconsolately to survey its fuzzy and plurivalent alternatives in hopes of finding something viable, however unlovely. Or one might dig in one's heels—recalcitrate, in a word—and accept the demarche as a lesson rather in the scope and limits of the notion of linguistic convention.

The first paragraph admits the floating hard boundary. He wants a line drawn but won't draw it. The second paragraph suggests that everything is convention and so why argue? In the 1970s they called that a copout.

Quine finds his molecule criterion awkward but not as "unlovely" as the fuzzy curve. I find the fuzzy curve accurate and complete. We can draw more than one curve but we never have to draw a line and face artificial borderline cases. Quine says bivalence is okay except at the boundary. But that is just the place we are fighting about. Why not smooth out the boundary with a curve? Here the very thing Quine argues against, multivalence or fuzz, can solve his bivalent problem. You can still keep some regions of 100% A and 100% NOT-A if it makes you happy. You do not have to draw the line between TALL and NOT-TALL at six feet. You can draw a ramp from five feet to six feet (Figure 9.14).

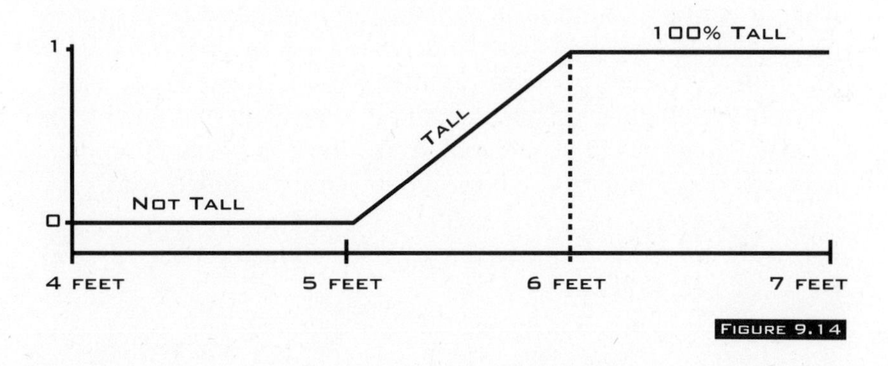

FIGURE 9.14

The steeper the ramp, the less fuzzy the term. You can even use tests and questionnaires and phone polls to help find the right shape. The ramp lets Quine and Aristotle keep their hard line. The ramp just lays the line on its side and lays it over the hard cases where no one wants to draw a hard line. Of course the ramp curve is still a fuzzy curve. It means that in the ramp part A AND NOT-A holds, the man is both tall and not, the hill is both a mountain and not, the molecule is both in the earth's atmosphere and not. Outside the ramp part A OR NOT-A still holds.

Quine and Aristotle and the rest might not agree with this. They may not want a ramp or any fuzzy curve. That's fine. Then they or their followers have to draw the hard lines and stick by them and defend them. As Quine says they have to get it right for each molecule. Good luck.

Then there are those who make the second bivalent claim, the scream of outrage. We like to think that such things do not belong in polite society and especially in science. And they don't belong there but society and science are full of them. In society we dismiss it as name calling or temper. We have learned to detect it and expect it and discount it when we find it. In science name calling may not be so easy to detect. Laymen and journalists fear science. Scientists work in narrow specialties and they fear the whole of science too. Science in this cultural sense can act like a church inquisition. Philosophers of science, like Quine, speak for science and tell us what is science and what is error or pseudo-science. No one can stand up to the key ideas of science for long. Once in a while it happens and there is a "paradigm shift" or scientific revolution, just as once in a while a gene mutation lasts in the gene pool. And bivalence is one of the deepest key ideas of science. It is so deep we take it for granted in science and in our attempts at scientific reasoning. Math is the language of science and it is bivalent and filled with precise categories of A AND not-A. Yet here is a short thin bald man down the hall from you in your Berkeley office who says that is all wrong. It is just a special case of gray.

In 1975 a reporter summarized Zadeh's place among his UC Berkeley peers this way:

> "Fuzzy theory is wrong, wrong and pernicious," says William Kahan, a professor of computer sciences and mathematics at Cal [UC Berkeley] whose Evans Hall office is a few doors down from Zadeh's. "I cannot think of any problem that could not be solved better by ordinary logic. . . . What Zadeh is saying is the same sort of thing as 'Technology got us into this mess and now it can't get us out.' Well, technology did not get us into this mess. Greed and weakness and ambivalence got us into this mess. What we need is more logical thinking, not less. The danger of fuzzy logic is that it will encourage the sort of imprecise thinking that has brought us so much trouble."

Kahan has not changed his position in the years since then. In 1990 a journalist asked me to comment on his recent claim that fuzzy logic is a license for loose thinking, that "Fuzzy logic" is "the cocaine of science." What can you say to that but thanks?

FROM FUZZY SETS TO SYSTEMS

Words stand for sets of things. These sets are fuzzy. The things belong to them to some degree. That means the things also belong to their opposites to some degree. So the Buddha's logic holds: A AND NOT-A holds to some degree. The man who is five feet seven inches is tall and not tall, maybe more tall than not. The man who is 30 years old is old and not old to some degree, maybe less old than not. A fuzzy set ties a curve or a point in a cube to a concept. The more the set A looks like its opposite NOT-A, the fuzzier it is and the greater its fuzzy entropy. The next step relates fuzzy sets.

So far we have looked at a fuzzy set like TALL as just one word in a sentence. Sentences have many words. They relate fuzzy sets and in this way we reason. Common sense. If the man is tall, he stands in the back. If it rains, you get wet. If the air is warm, turn up the air conditioner.

A group of sentences gives a fuzzy system. In the simplest case someone tells you the sentences. They give you their expert advice or rules of thumb or common sense. A fuzzy system takes this knowledge and models the world with it. The system may control a washing machine or give buy and sell signals on stocks and bonds. In the adaptive case the fuzzy system comes up with its own sentences. It not only reasons with its knowledge, it also grows its knowledge from data. It learns from experience. That is an adaptive fuzzy system. In all cases the building blocks are fuzzy sets.

10

Fuzzy Systems

TO: The Secretary of State, Washington, D.C.
FROM: U.S. EMBASSY IN TOKYO, JAPAN

Japanese government, industrial, commercial, and academic institutions are busily studying fuzzy-logic theory and using fuzzy logic in a variety of applications. The government research project is led by the Science and Technology Agency's five-year research project consisting of nineteen separate programs (e.g., simulating global air pollution, predicting earthquakes, modeling plant growth). The Japanese industrial effort is showcased by the MITI-authorized Laboratory for International Fuzzy Engineering (LIFE), established by 48 Japanese companies to strengthen academic-industrial ties. Some of the applications which LIFE is working toward are a nuclear-power-plant control system and a prototype fuzzy computer. . . . Japanese fuzzy-system researchers expect that fuzzy logic will enable the development of computer systems which adjust to people, rather than the reverse.

MEMORANDUM R-120608Z
MARCH 1990

F*uzzy logic is a concept derived from the branch of mathematical theory of fuzzy sets. Unlike the basic Aristotelian theory that* *recognizes statements as only "true" or "false," or "1" or "0" as represented in digital computers, fuzzy logic is capable of expressing linguistic terms such as "maybe false" or "sort of true." In general, fuzzy logic, when applied to computers, allows them to emulate the human reasoning process, quantify imprecise information, make decisions based on vague and incomplete data, yet by applying a "defuzzification" process, arrive at definite conclusions.*

FUZZY LOGIC: A KEY TECHNOLOGY FOR FUTURE COMPETITIVENESS
U.S. DEPARTMENT OF COMMERCE, NOVEMBER 1991

***T**he way of the warrior is resolute acceptance of death.*

MIYAMOTO MUSASHI
THE BOOK OF FIVE RINGS

***B**usiness is war.*

NEW JAPANESE PROVERB

How do you make a machine smart? Make it FAT. Put some FAT in it. That's how you put the Buddha's logic in machines. In theory a FAT enough machine can model any process. A FAT system can always turn inputs to outputs and turn causes to effects and turn questions to answers.

FAT stands for *Fuzzy Approximation Theorem*. In 1990 I proved that a fuzzy system can model or approximate *any* system.* The FAT theorem shows why fuzzy systems have raised the machine IQ of camcorders and car transmissions and helicopter stabilizers and why future fuzzy systems can raise MIQs even higher.

---

*A detailed but early paper is my "Fuzzy Systems as Universal Approximators," *Proceedings of the First IEEE Conference on Fuzzy Systems (FUZZ-92)*, 1153–62 San Diego, March 1992.

The FAT idea has a simple geometry: Cover a curve with patches (Figure 10.1). Below we will look at the FAT idea in detail. For now you can think of it this way. Each piece of human knowledge, each rule of the form *IF this THEN that*, defines a patch. A fuzzy system is just a big bunch of fuzzy if-then rules—so it is just a big bunch of patches. All the rules define patches that try to cover some wiggly curve. The better the patches cover the curve, the smarter the system. More knowledge means more rules. More rules means more patches and a better covering. The more uncertain the rules, the bigger the patches. The less fuzzy the rules, the smaller the patches. If the rules are so precise they are not fuzzy, then the patches collapse to points and they don't cover much of anything.

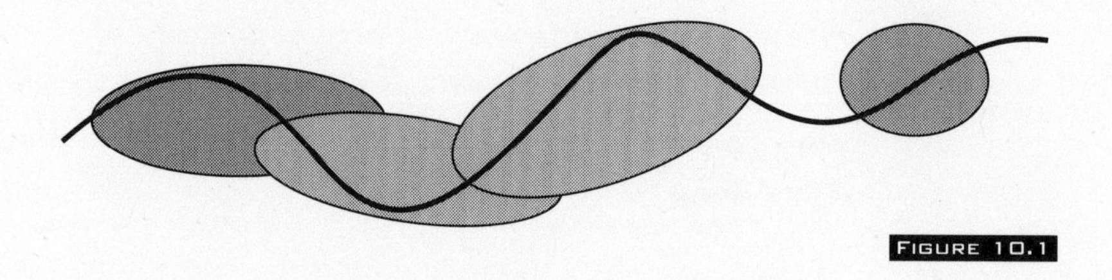

FIGURE 10.1

This chapter looks at how fuzzy systems work and at how Japan works with fuzzy systems. How fuzzy systems work comes down to how they reason with FAT. We start with a single rule or patch and then look at many patches that work together as a fuzzy system. The next chapter looks at how fuzzy systems can learn their own patches.

KNOWLEDGE AS RULES

How do you reason? You want to play golf on Saturday or Sunday and you don't want to get wet when you play. The news says there is a good chance it will rain on Saturday but only a slight chance it will rain on Sunday. You reason that you should play golf on Sunday. But how do you reach this answer?

You reach it with rules. Rules associate ideas. They relate one thing or event or process to another thing or event or process.

In natural and computer languages rules have the form of if-then statements. If it rains, you get wet. If you get wet, you can't play golf. It will rain on Saturday. So you can't play golf on Saturday. It won't rain on Sunday. If you can't play golf on Saturday and if it won't rain on Sunday, you can play golf on Sunday. So you play golf on Sunday.

Knowledge as rules goes back to at least Aristotle. Leibniz dreamed of a symbolic logic and *ars combinatoria* or computer system that could put all our rules in symbols and reason from them to reach math truths and daily facts. Today computer scientists have built the field of "artificial intelligence" or AI on the belief that knowledge is rules *and* that you can write down rules in the black-and-white language of computers and symbolic logic. After over 30 years of research and billions of dollars in funding AI has so far not produced smart machines or smart products. The AI fans claim it's only because they can't yet put enough rules in computers. AI "expert systems" use 100 to 1,000 or so bivalent rules. Some AI experts claim we will not see "real" intelligence in AI systems until they use 100,000 rules. We critics call that throwing more rules at the problem.

Fuzzy researchers have built hundreds of smart machines and we think we know why the AI folks have failed. Yes, you need rules. No, you don't need a lot of rules for many tasks. You need *fuzzy* rules.

If it rains, you get wet. We mean a lot with that rule. If it rains a little, you get wet a little. If it rains a lot, you get wet a lot. The noun *rain* stands for a fuzzy set. Rain can spit or drizzle or shower or drench. *Little* and *a lot* stand for fuzzy subsets of rain. All rain falls a little or a lot. It's a matter of degree. That's what computer or AI rules miss. Every term in one of our rules is fuzzy. Every term is vague, hazy, inexact, sloppy, shot through with exceptions. One human rule covers all these cases. An AI or computer rule covers just one precise case that someone picked on a good day or a bad day as they stroked the keys on the computer.

A fuzzy rule relates fuzzy sets. IF X is A, THEN Y is B. A and B are fuzzy sets. They are fuzzy subsets of X and Y. If the rain is heavy, you get very wet. If the car slows a little, push the gas pedal down a little. If the car wheels turn to the left a lot, turn the steering wheel to the right a lot.

The AI folks like to call rules "heuristics" or rules of thumb.

They are what an expert says to explain what he does. They are *common sense*. And yet what we sense is not common to us. The world hits us with signals and we sense them through the unique eyes and ears and tongue and skin that our DNA gives us. Sense is common only to some degree even if we see and hear and taste and touch the same things. And the world hits each of us with a unique set of signals. That's why every word stands for a different fuzzy set of things to each of us. A child's idea of *large* or *slow* or *fair* changes as she grows older.

The AI crowd missed that. They went for rules but scrubbed all the gray out of them. The first AI researchers worked at the tail end of the logical positivism of the twentieth century. They were eager to apply the black-and-white symbolic logic of their teachers. They applied it and made wild claims and promises to the press. They took state funds and defense-buildup funds and set up their own classes and conferences and power networks. And they did not produce a single commercial product that you can point to or use in your home or car or office. They beat up on fuzzy logic harder than any other group because they had the most to lose from it and the fastest. Fuzzy logic broke the AI monopoly on machine intelligence. Then fuzzy logic went on to work in the real world.

I remember when I met Lotfi Zadeh at the 1984 AI conference in Austin, Texas, the city that had just helped pay for the giant AI computer consortium MCC (Microelectronics and Computer Technology Corporation) to "keep up" with the Japanese AI effort that later failed. Lotfi sat on a panel debate on uncertainty in AI with four young AI researchers. They refused to address him or rebut his claims except with a wave of their hands to Lotfi and to the audience. I thought old Lotfi would tower over these young men but they laughed him down. Part of the problem was that Lotfi did not have any math to show them or could not at the time cite any commercial fuzzy products. The other part of the problem was that Lotfi did not share their faith that bivalent logic and rules made sense in the real world. He said that approach "put computers on a short leash" and was "like trying to dance a jig in a suit of armor." The AI crowd laughed at the jig joke and at Lotfi when he tried to answer one of the speakers but the speaker would not let him. Those days ended when Japan began the fuzzy race for higher machine IQ and when at about the same time the fall of the Soviet Union ended the easy defense money for AI research.

The fuzzy view was just common sense. It was daring and novel at the time because you first had to get your university degrees in the old black-and-white school and then doubt that school and rediscover what any layman could have told you about common sense—it's vague and fuzzy and hard to pin down in words or numbers.

That still does not persuade hard scientists. You have to show them the math before they will accept it. And the best way to do that is to show them pictures. Show them geometry. Show them a fuzzy rule as a fuzzy patch.

RULES AS PATCHES

A fuzzy rule defines a fuzzy patch. Patches and shades of gray are the two key ideas of fuzzy logic. They tie common sense to simple geometry and help get the knowledge out of our heads and onto paper and into computers. Patches come right out of how you build a fuzzy system. So let's build one.

You build a fuzzy system in three steps. First, you pick the nouns or "variables." Call these X and Y. X is the input to the system. Y is the output. Input, output. If X, then Y. Cause, effect. Stimulus, response. Question, answer. Say we want to control an air conditioner. Let X be temperature in degrees Fahrenheit. Let Y be the change in the motor speed of an air conditioner. We want the motor to speed up when it gets hot and slow down when it gets cold.

Second, you pick the fuzzy sets. We define fuzzy subsets of the nouns X and Y. Say the five fuzzy sets on X are COLD, COOL, JUST RIGHT, WARM, and HOT. We draw these as curves or triangles as we drew the fuzzy number ZERO in the last chapter. See Figure 10.2. We could draw the curves in different places or use bell curves or trapezoids or anything else instead of triangles. In practice triangles work just fine. It's a matter of common sense and engineering judgment. I have drawn the COLD and HOT sets as "half" triangles. These still count as "full" triangles because you can always draw in the missing halves. It won't matter since they lie outside the temperature range.

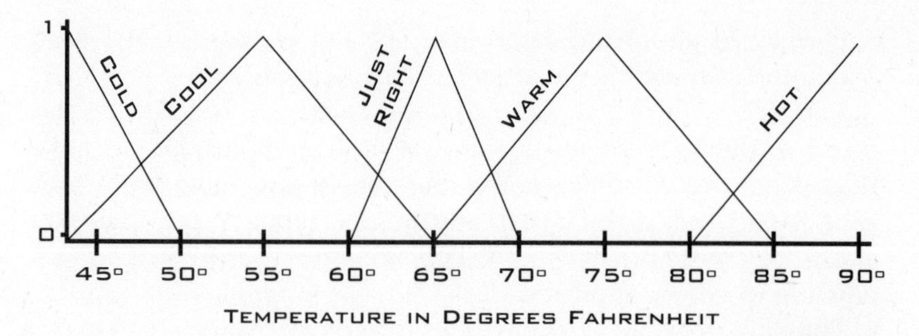

FIGURE 10.2

I have also drawn some sets wider than others. This is control common sense. The widest sets are least important. They give only rough control. For fine control you want thin sets. The set JUST RIGHT covers the desired temperatures near 65°. We want the control system to bring us quickly from hot or cold and carefully zoom in on what feels just right. This helps avoid the "overshoot" and "undershoot" that you feel with car or house air conditioners or with most black-and-white controllers, from cruise controller to microwave ovens.

Next we draw the five fuzzy sets STOP, SLOW, MEDIUM, FAST, and BLAST on the motor speed (Figure 10.3). I put motor speed in numbers from 0 to 100. These numbers can stand for the electric current to the motor or the r.p.m. of the motor or a fan. I have drawn MEDIUM thin for fine control near the desired speed.

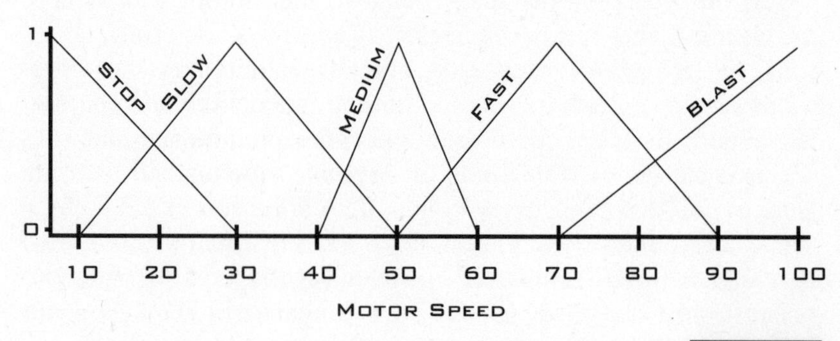

FIGURE 10.3

Third, you pick the fuzzy rules. This step associates motor-speed sets with temperature sets. We have to assign a motor-speed set to each temperature set. Start with COLD. We want the motor to shut off if the air is too cold. The air conditioner should change the motor speed so as to keep the temperature just right. So we have our first rule: IF X is COLD, THEN Y is STOP. The motor should run a little when the air gets cool: IF X is COOL, THEN Y is SLOW. The motor should run *medium* when the air feels just right: If X is JUST RIGHT, THEN Y is MEDIUM. It should speed up when the air gets warm: IF X is WARM, THEN Y is FAST. And the motor should blast like hell when the air gets hot: IF X is HOT, THEN Y is BLAST. That gives five rules in English:

Rule 1: If the temperature is cold, the motor speed stops.
Rule 2: If the temperature is cool, the motor speed slows.
Rule 3: If the temperature is just right, the motor speed is medium.
Rule 4: If the temperature is warm, the motor speed is fast.
Rule 5: If the temperature is hot, the motor speed blasts.

Those rules are just common sense. They are fuzzy rules because the terms "cold" and "fast" and the like are matters of degree and stand for fuzzy sets. We tied them to fuzzy sets defined on numbers. That ties words to math. And where there is math there is geometry. That leads to a new question.

What do fuzzy rules *look like*? They look like patches. What do you get when you cross a triangle with a second triangle? A triangle? No. You get a patch. If the sets are warm and fast, you get rule 4: If the temperature is warm, the motor speed is fast. And the rule is a patch (see Figure 10.4).

In math you call this patch the math product of two triangles. What is the math product of two line segments? A square or rectangle. Let's look at math products for a minute to build your intuition as you do in the field of algebraic topology. What is the math product of three line segments? A solid box or cube. That also equals the math product of a solid square and a line segment. Now here's a hard one: What is the math product of a circle and a solid disk? A bagel or doughnut or smoke ring or "torus." You can see this when you swing a tennis racket in a

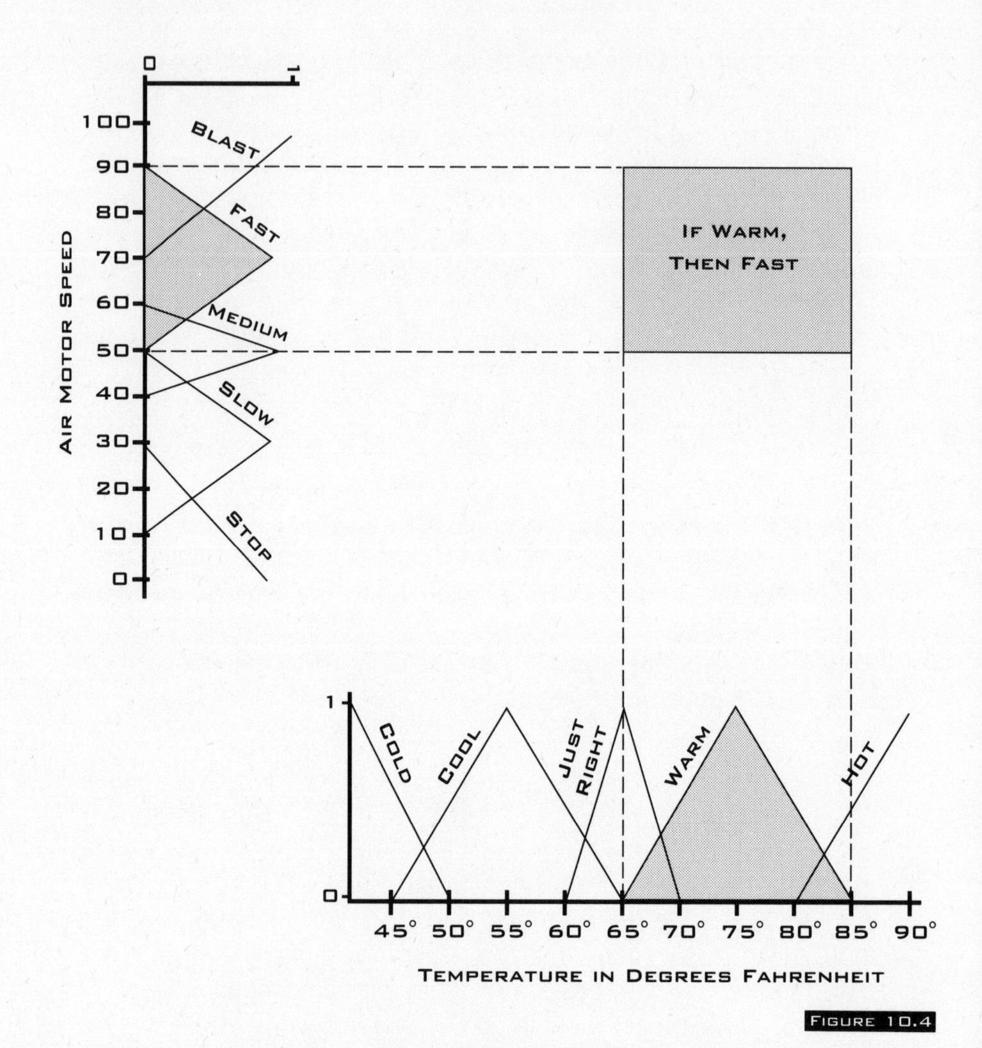

FIGURE 10.4

circle. If you cross just a circle with a circle, you get the skin or surface of a torus—a bagel with no calories. If you cross a spike or point with a spike you get a point. So AI or bivalent rules define points. They define patches that have shrunk to points. And it's hard to get FAT with a few points.

I have drawn the rule patch as a rectangle. A rectangle comes from the math product of two line pieces. The first line piece is the base of the WARM triangle. The second line piece is the base of the FAST triangle. That's all I have drawn.

The real rule patch is a 3-D figure and hard to draw on the

page. It looks like a tent with four sloped poles on the sides and one tall pole sticking straight up in the center. Each point on the tent floor is a pair of X-Y numbers, a temperature degree and a motor speed. The tent height tells you the degree to which the number pair belongs to the rule patch. Only the center point belongs 100%. The other points belong to lesser degrees. Outside the tent floor all the other points belong 0%. So the patches are fuzzy.

We write the math product of WARM and FAST as WARM × FAST just as you would write the product of two numbers. The same goes for the other four rules. We write RULE 1 as COLD × STOP, RULE 2 as COOL × SLOW, RULE 3 as JUST RIGHT × MEDIUM, RULE 5 as HOT × BLAST. We start with words or ideas and end with geometry. First we come up with common-sense rules in words. Then we tie these to fuzzy sets. That gives us the names of patches like WARM × FAST and HOT × BLAST. Then we draw the patches.

That means we can draw or see the entire fuzzy system as five patches that overlap (Figure 10.5).

Now the system gets FAT. Look at how the five patches hang together. They cover something. They cover a line that runs from the lower left to the upper right. They cover many lines. They cover a linear system. Double the input and you double the output. Triple the input and you triple the output. And so on. Our fuzzy system is nonlinear but it approximates a linear system.

I picked a "toy" or simple control problem and built a toy fuzzy system to solve it. We saw earlier in Chapter 4, The Ways of Paradox, that linear systems are the toy problems of science and yet most scientists treat real systems as if they were linear. That's because we know so little math and our brains are so small and we guess so poorly at the cold gray unknown nonlinear world out there. Fuzzy systems let us guess a little better and that has made all the difference for smart machines. Fuzzy systems let us guess at the nonlinear world and yet do not make us write down a math model of the world. We wrote down no equation for our fuzzy air conditioner. That's where fuzzy systems break with old science. The technical term for it is *model-free* estimation or approximation. You do it every time you back up your car or catch a fast ball or look at a TV image and see something in your brain.

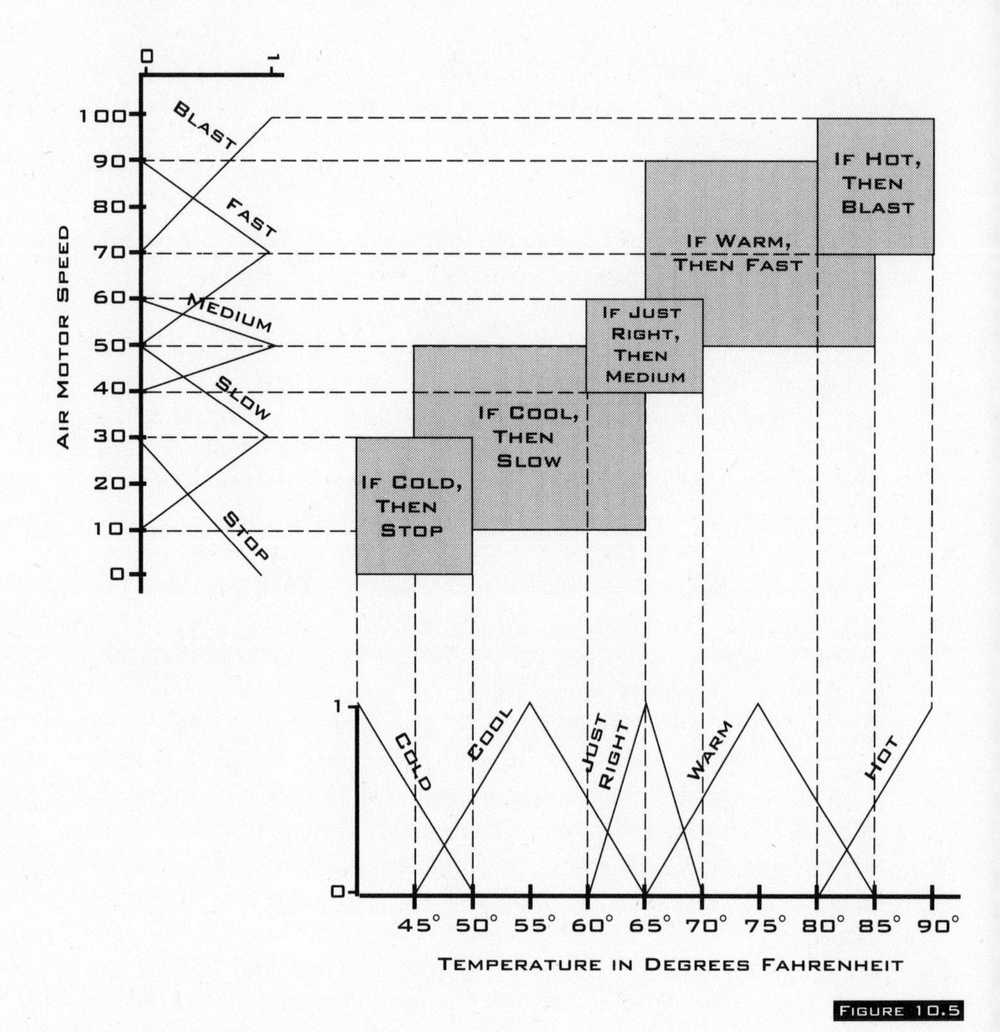

FIGURE 10.5

You can see this point with figures 10.6 or 10.7. The idea is that fuzzy patches can cover any curve that wiggles. A wiggle means the system is nonlinear. A straight curve means it is linear. We will return to how a fuzzy system comes up with a nonfuzzy answer or motor speed to an exact measured temperature value. While we have this system patch picture I want to pause here and get to the meat of fuzzy systems. And the meat is FAT.

THE FAT THEOREM

FAT stands for *Fuzzy Approximation Theorem*. So "FAT Theorem" is redundant but sounds better than "FA Theorem" so I will use it. The FAT Theorem says you can always cover a curve with a finite number of fuzzy patches. You cover the curve just as you would block out a black line drawn on a sheet of dough with a cookie cutter. To cover the curve you let the cuts overlap. The smaller the cookie cutter, the finer the cover but the more times you have to cut.

Sloppy rules give big patches. Fine rules give small patches. The less you know about a problem, the sloppier your rules. You tend to use fewer fuzzy sets and they stand for more things and cases. A simple air conditioner may use only COLD, JUST RIGHT, and HOT to describe temperature. These fuzzy sets define wider triangles that overlap. The system above used more fuzzy sets over the same range of temperature values. It used COLD, COOL, JUST RIGHT, WARM, and HOT. These triangles had less width. We could add more fuzzy sets and draw them with thinner triangles. In the limiting case the width shrinks to zero and the triangles become spikes. The spikes define regular numbers. And there are infinitely many spikes between any two temperature values. The fuzzy sets lose their fuzziness as the triangle widths shrink. Along the way they lose their common-sense meaning. It's hard to work with lots of small triangles but they are more precise than wide ones. The moral: *You pay for precision*. Precision, like clean air, is not a free good.

The rule patches shrink as the triangles get narrower. A few wide fuzzy sets may give a rough cover of the nonlinear system (Figure 10.6).

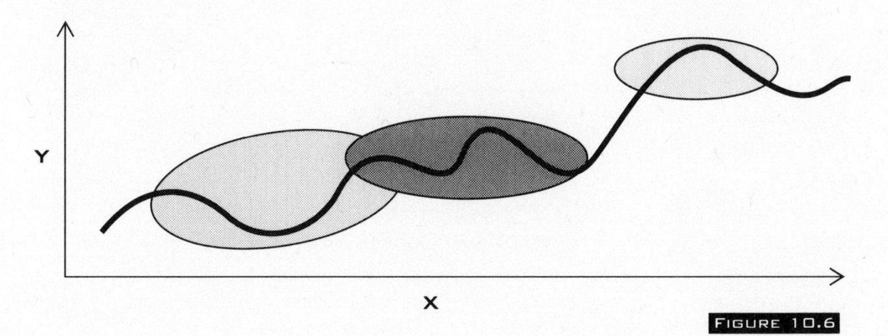

FIGURE 10.6

I have drawn the system as a curve that wiggles from left to right. A system is anything that maps inputs to outputs, X things to Y things, causes to effects. A nonlinear system has a curve that wiggles. A continuous system has a curve with no gaps or tears in it. The FAT Theorem says a fuzzy system can approximate a continuous system (on a "compact" or closed and bounded set) as closely as you please. This includes all systems studied in calculus. It also includes almost all systems studied in science.

You can see the idea behind the FAT Theorem as you shrink the patches and add more of them. This gives a finer cover (Figure 10.7). Keep this up for a while. Shrink and add patches. Shrink and add more patches. The cover gets finer and finer and uses only a finite number of rules. The idea is simple enough but the proof takes some work. It uses topology and measure theory. Topology deals with smooth changes in surfaces as when you roll a flat piece of paper into a tube. Measure theory deals with area as when you cut a table top in four pieces and find that the area of the four pieces equals the area of the uncut table top. Differential topology lets you take the FAT Theorem another step. With it you can show that the FAT Theorem still holds if the curve moves and wiggles. If the curve does not wiggle too fast, you can always move and shrink or grow the patches to keep up with it. In control theory this means fuzzy systems can model *dynamical* systems that change with time.

The FAT Theorem shows why fuzzy systems work. Common sense gives patches and these cover a system you want to control. In practice no one knows how to write down the

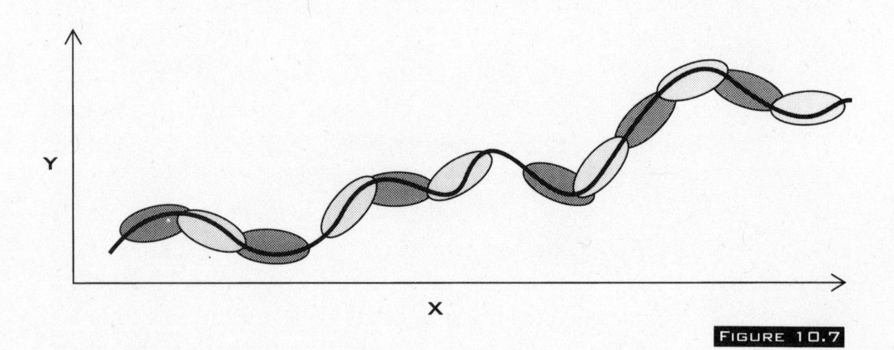

FIGURE 10.7

equations for most systems. They are too nonlinear. Fuzzy systems guess at the equations for you. They let you express what you know in a nonmath language. It took me some time to see this. Fuzzy systems seemed too good to be true. They let novices build control systems that beat the best math models of control theory. You can see it in the patches.

It was not always so. We had only a seat-of-the-pants feel for why fuzzy systems should work. Consider this quote from Professor Ebrahim Mamdani of Queen Mary College in London. Mamdani built the first fuzzy system in the early 1970s to control a steam engine and later built and field-tested the first fuzzy traffic lights. In 1978 he described his work this way:

> The basic idea behind this approach was to incorporate the "experience" of a human process operator in the design of the controller. From a set of linguistic rules which describe the operator's control strategy, a control algorithm [system of fuzzy rules] is constructed where the words are defined as fuzzy sets. The main advantages of this approach seem to be the possibility of implementing "rules of thumb" experience, intuition, heuristics, and the fact that it *does not need a model of the process.* (Italics added.)

The key is no math model. Model-free estimation. *Model freedom.* If you have a math model, fine. But where do you find one? You find good math models only in textbooks and classrooms. They are toy answers to toy problems. The real world pays no attention to most of them. So how do you know how well your math guess fits Nature's process? You don't and you never can. You have to ask God or No God or Nature and no one knows how to do that. Short of that we guess and test and guess again. Scientists often respect math more than truth and they do not mention that a math guess is no less a guess than a guess in everyday language. At least a word guess does not claim to be more than a guess and gains nothing from the fact that we have stated it in words. A math guess has more dignity the less math you know. Most math guesses are contrived and brittle and change in big ways if you change only a small value in them. Man has walked Earth for at least a million years and has just started to think in math and is not good at it. Fuzzy systems let us model systems in words. Cro-Magnon and Neanderthal man had enough common sense in their heads and maybe in their words to feed fuzzy systems.

The FAT Theorem says more than this. It says you can get rid of all the books on physics and chemistry and biology and economics and replace them with new books that have fuzzy systems where equations used to be. In theory we can translate all the equations into rule patches. We need not limit fuzzy systems to the control of robots and household gadgets and car parts. They approximate systems. You can use them in physics or communications or neuroscience. I have worked with graduate students on fuzzy smart cars and communication systems and fuzzy "virtual reality" systems that model the undersea world of dolphins as they feed for fish and flee sharks. Students in my neural-fuzzy class at USC compete for a $1,000 prize that some company donates and have built or simulated smart systems that lock a telescope on a star as the Earth rotates, match couples for dates, open a cat door for your cat only, chlorinate a pool, land a lunar probe, bench-press a weight for full burn, bet on stocks or blackjack hands, guide a submarine, advise you how to seduce a date, cool injection-molded plastic, pass on an LA freeway, or use a robot arm to grab an astronaut, make an IV tube smarter, adapt a heart pacemaker, wipe a windshield, plow and fertilize and plant a field, incubate a premature baby, run a ski lift or sprinkler system, humidify a greenhouse, or steer an antenna, to list but a few. Most students walked into my class with no knowledge of fuzzy systems and in less than three months had built, tested, compared, polished, videotaped, and shown the class a fuzzy system that worked. A few have gone on and patented their projects.

You can use fuzzy systems anyplace we today use brains. And you can use them in places where even brains don't work. Professor Michio Sugeno of the Tokyo Institute of Technology has built a fuzzy system that can stabilize a helicopter in flight when it loses a rotor blade. No human can do that and no known math model can do that. Sugeno's system uses about 100 rules. He has tested it on a three-meter model and wants to test it on the real thing. Sugeno added voice control so a pilot can guide the helicopter with terms like "up," "down," "hover," "left," and "right."

Sugeno's work differs in kind from the work of control engineers in the U.S. and Europe. Sugeno built a real system and tested many math ideas and hypotheses on it. I think he is the best fuzzy-math theorist in Japan and he knows as much math as anyone in science. In the U.S. or Europe theorists like me do not build things. We write papers and prove theorems and argue math. We look down on those who dirty their hands with simulations or, worse, who build and test real systems. The creed is "That's what graduate students are for." Scientists respect what they do not understand. And few know or understand much math and the next generation always learns more math. Yet here is Michio Sugeno, former Marxist labor radical in his student days and later consultant to the largest firms in Japan and thus in the world and the quiet man behind the big fuzzy consortium LIFE in Yokohama, building a big toy that flies to test ideas we do not even know how to state in math. That's real engineering and the way of the future—at least in Japan.

Knowledge is rules. Rules are patches. Patches cover the curve of a system. But that does not tell you how to convert 65° into a motor speed. Patches and the FAT Theorem show you why a fuzzy system works. We now look at how they work.

FUZZY ASSOCIATIVE MEMORY: FIRE ALL RULES

In a fuzzy system which rule "fires" or activates at which time? They all fire all the time. They fire in *parallel.* And all rules fire to some degree. Most fire to zero degree. They fire *partially.* Parallel and partially. That's how *associative* memory works. The result is a fuzzy weighted average.

Look at our fuzzy air conditioner. Say the thermometer reads 65°. This is the input to the system of five rules. We need to convert 65° to a motor speed. The input 65° can belong in some degree to all five fuzzy sets, COLD, COOL, JUST RIGHT, WARM, and HOT (Figure 10.8). We look at the triangles and see that it belongs 100% to JUST RIGHT and 0% to all the rest. That means RULE 3 fires 100%. If the temperature is just right, then the motor speed is medium. RULE 3 is the patch JUST RIGHT × MEDIUM. So we get as output the set MEDIUM (Figure 10.9).

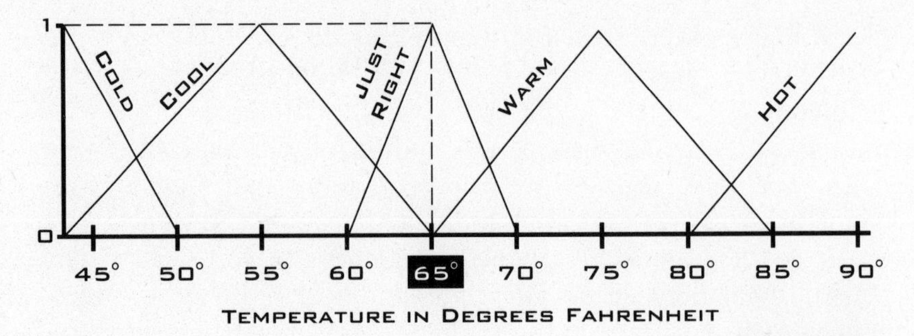

FIGURE 10.8

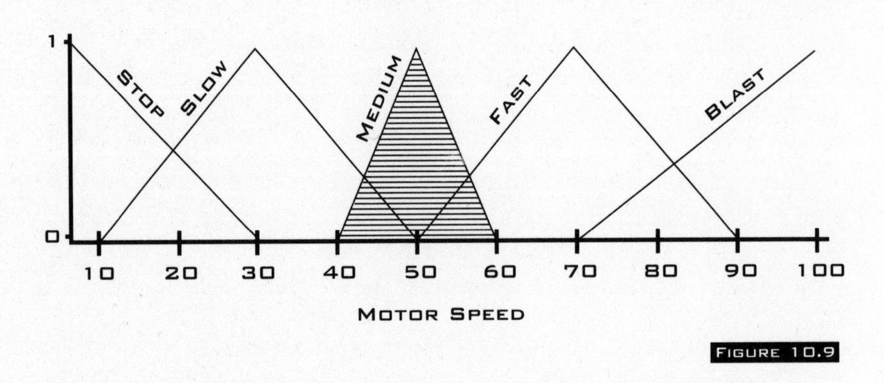

FIGURE 10.9

You can't hand a fuzzy set to a motor. You have to tell the motor a speed. You have to give it a number. A triangle is not a number. But it has an *average* number. Since the MEDIUM triangle is symmetric the average is just the center of the base of the triangle, the motor speed 50. This average is the *centroid* or center of mass of the output set. We call this *defuzzification.* You defuzzify a set when you replace it with a number or centroid.

A fuzzy chip walks through this process millions of times per second in FLIPS or *fuzzy logical inferences per second.* A fuzzy chip takes an input number and compares it to all input fuzzy sets and gets output sets and converts them to an output number and then repeats these steps over and over. We just dealt with the simplest case, when the input fires one rule, and showed how 65° leads to motor speed 50.

Now say the temperature falls to 63°. We look at the set

triangles and see that this input belongs 80% to JUST RIGHT and belongs 15% to COOL and belongs 0% to the other sets (Figure 10.10).

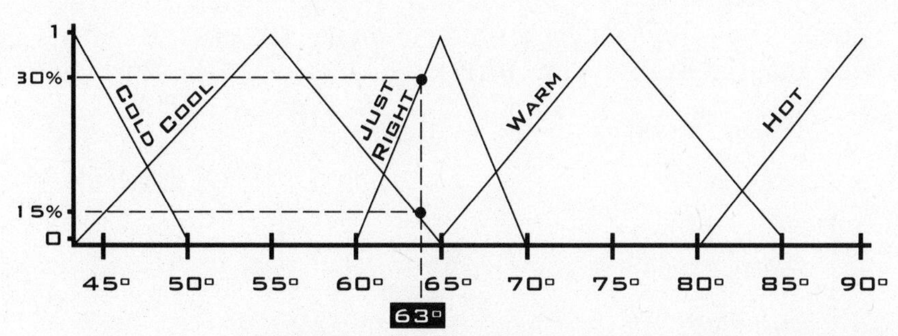

FIGURE 10.10

So two rules fire but each fires only to some degree. RULE 3 fires only 80%: If the temperature is just right, then the motor speed is medium. But now RULE 2 fires 15%: If the temperature is cool, then the motor speed is slow. This gives two output sets or triangles. But we shrink the MEDIUM set to 80% of its height and shrink the SLOW set to 15% of its height (Figure 10.11).

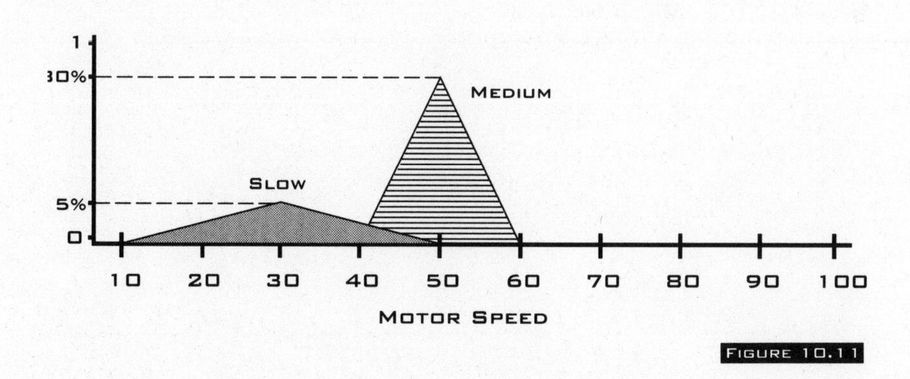

FIGURE 10.11

What do you do with two triangles that overlap? Good question. Old fuzzy systems took the outline or silhouette of the two triangles to form the final output set. I showed that if you do this with too many sets you get back a flat line and that always has the same average value. To fix this I showed that you should *add* the two triangles to get the output set. I call this an *additive fuzzy system*. Most fuzzy systems today are additive and that's a good thing since for them the FAT Theorem holds. If you add enough sand hills you get a sand mountain. In our case you get back a jagged peak (Figure 10.12).

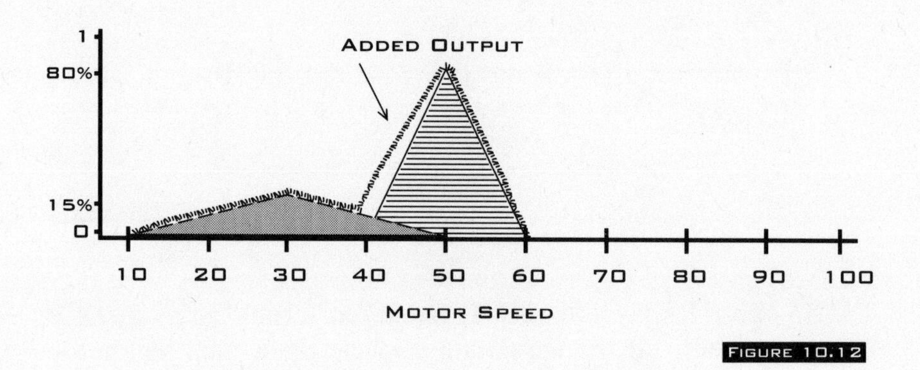

FIGURE 10.12

We now "defuzzify" this set or take its average value to find the output motor speed at about 42 (Figure 10.13). The fuzzy system slows to let the air warm up a little.

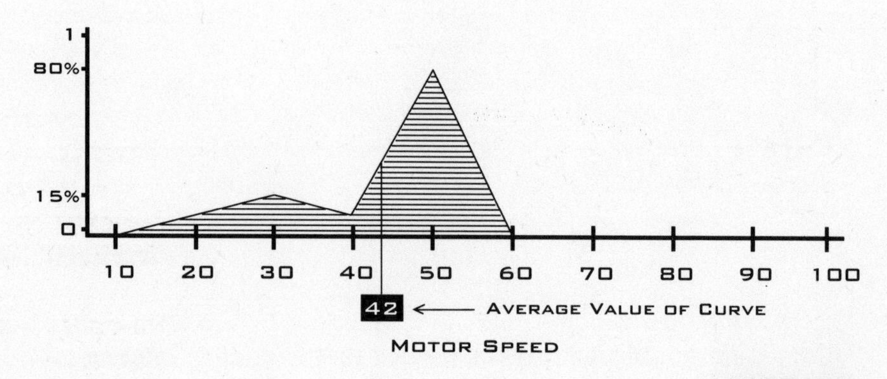

FIGURE 10.13

We can repeat these steps for any temperature value. In each case an input number gives one output number. If we change the fuzzy sets, we change the output numbers. If we add or delete rules or patches, we change the output numbers.

A chart shows how a fuzzy system "reasons" or maps inputs to outputs (Figure 10.14). Each rule has the form IF A, THEN B. This is short for *IF X is A, THEN Y is B*. The input data x fires the A part of each rule to some degree. That gives B′ or the B part of the rule to some degree. The system adds the B′ sets to get a final output set B and then takes the average of B to get the output number y. The output y is a fuzzy weighted average.

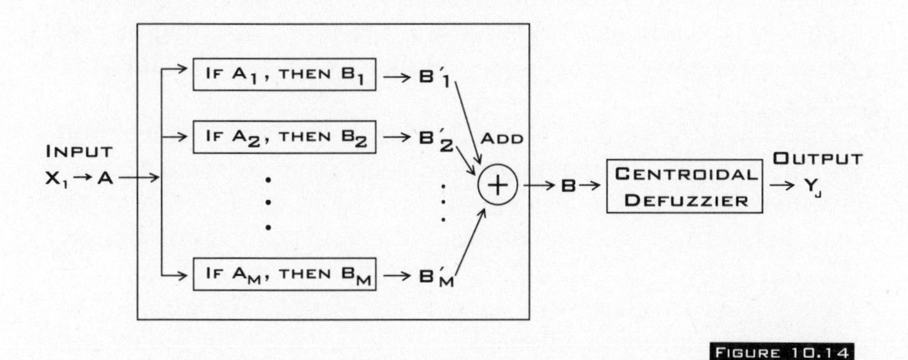

FIGURE 10.14

All systems turn inputs to outputs. Now you know how a fuzzy system or fuzzy chip or fuzzy software does it. The Sendai subway uses 59 fuzzy rules. Pocket cameras use 10 or so rules to control autofocus. The Yamaichi stock trader uses about 1,000 rules. But they all obey this chart.

I use the term FAM or *fuzzy associative memory* to describe how a fuzzy system works. Parallel and partially. Fire all rules to some degree. Computers use *direct* memory. They store a thing at an address or cell and look it up a cell at a time. Associative memory searches or fires the whole memory just as we search our whole past when we try to place a familiar face in a crowd. For math reasons I also use FAM to refer to a single rule in the FAM chart (Figure 10.14). That is a long story and you could write a book about it and I did. You don't need the book to see why. A FAM system with only one rule is still a FAM. So a rule is a FAM.

The FAM chart is a model of how we reason. You can defend it with how well a fuzzy chip works. You can also defend it by analogy. I want to show how a FAM system comes close to how we decide what to do in our own lives and how a law judge decides how to rule in our lives. After that we will look at real FAM systems and the folks who make them and fight over them.

FUZZY DECISIONS IN LIFE:
Fuzzy Weighted Averages

How do you decide to do or not do? What do you do next? You get a new job offer. Do you take it? You get asked on a date or to marry. Do you say yes? You want to buy a new dog or new car or new house. Which do you pick? Can you explain your choice?

The FAM view is that you pick a fuzzy weighted average. You add up a lot of things and weight each thing to some degree. Then you go with the average or "centroid" or center of mass. You do not solve math equations or draw rule patches on a page. You do it by feel. You feel or intuit the center of mass. It pulls you or inclines you. After a life of this you are good at going with your mind's flow. This picks answers but it does not explain them.

Suppose you get a new job offer. And you take it. What do you tell your friends when they ask why you took it? You will say a lot of things but each thing will be only part of the answer: Better pay. Liked the people. More room for growth and advancement. More interesting work. Better benefits. Bigger office and better view. Shorter drive to work. Can't stand my old boss. You can *defend* your choice with these reasons. But you cannot fully *explain* it. Each reason you give is a matter of degree. Each has a fuzzy or gray weight. How much better is your new pay than your old pay? How much more do you like the people? How much shorter is the drive? No one reason throws the decision. They all add up to the decision. A fuzzy weighted average.

Now look at two gardens in summer. Which do you find more beautiful? The first is a New England garden with flowers and herbs: violets, morning glories, purple carpets of alyssum, mounds of rainbow zinnias and white daisies, and clumps of lavender, green spearmint, red clover, dill, thyme, goldenseal, and wild ginger. The second garden is a Japanese rock garden such as you might find next to a Zen temple in Kyoto. Green mosses grow on the bumpy roots and smooth black bark of Japanese maples with finely serrated green leaves. Monks have raked fine white gravel in spirals around big mossy rocks. Orange and red carp swim in the still pool that borders the rock garden on one side. On the other side lie more stones and moss and a bamboo fence tied with rope. Behind it stands a green bamboo forest.

Both gardens are package deals. They both look good and smell good. And the beauty of the whole depends in some way on the beauty of the parts. How do you vote? You pick the one that feels best. In defense you can cite lists of what you like and do not like. Each is a matter of degree as is the clash of the white daisies with tall bunches of lavender or the symmetry of the raked gravel. A fuzzy weighted average can approximate your choice. That's what the FAT Theorem says. Your choice depends on complex and nonlinear processing in your brain. At a high level it looks like a fuzzy weighted average, or the center of mass of a curve.

When I talk about weighted fuzzy averages to some of my fuzzy friends they don't like it. Probability theory has staked out the term "weighted average" or the so-called mathematical "expectation." I showed that these weighted averages lead to an optimal probability decision, a so-called conditional mean. You have to give a probability interpretation of a fuzzy set to get the math to go through but you can always do that. You can view the set of COOL temperatures as an infinite list of probabilities—the probability that the air is cool at each temperature. Some of my fuzzy friends thought I had lost the faith when I first talked about this. One writer friend even called it "Bart's apostasy." But optimality is no vice. Fuzzy systems make weighted fuzzy averages. That is a fact and from a math point of view a blessing. What really counts with fuzzy systems, the real value added, is the tie between words and sets and between knowledge and patches. Certain knowledge, small patch. Uncertain knowledge, big patch. No math model. That's the advance. The weighted averages and the rest are details that sit on top of the patch structure. And we seem to make weighted averages when we make a decision. Don't we?

A lot of people don't think so. They have a different answer from the FAM answer. The bivalent answer. It comes from Aristotle and AI gurus and the class projects of MBA grads. The logical answer. Chains of logic. A branch of a decision tree. If A, then B. If C, then D. If B and D, then E. Each statement holds all or none. This is an answer by rule book and works with the letter of the law and not the spirit of it. In law they call it a *per se* decision. They call the fuzzy spirit of the law the *rule by reason*. The split comes up when you try to replace a court judge with a computer.

FUZZY SYSTEMS AS JUDGES: RULES *VS.* PRINCIPLES

How does a court judge decide a case? Does she follow the letter or the spirit of the law? Does she open a rule book and match "facts" to the if-parts of rules and then read off the verdict or sentence? Or does she use the rule of reason and match fuzzy facts to fuzzy case precedents and end up with what looks like a fuzzy weighted average?

The letter and spirit of the law arise from the split between rules and principles. Rules are precise, black and white, all or none. They have no exceptions. You get kicked out or fired or fined or jailed if you break them. We post rules on signs or send them in memos or print them on contest forms and drivers' licenses: No one under 17 admitted without parent or guardian. Must be at least 35 years old to run for President of the United States. $1,000 fine for littering. Mall hours 10 A.M. to 9 P.M. Three weeks of vacation per year. Maximum speed 65 m.p.h. Right turn only.

Principles are vague and abstract and full of exceptions. They change slowly with time as culture evolves. The U.S. legal system got most of its principles from medieval British common law and the Magna Charta: Innocent until proven guilty. Freedom of contract. Equality before the law. Full disclosure. Buyer beware. You cannot profit from a crime. You cannot challenge a contract if you act on it. Buyer's remorse.

Law as rules. That view still holds sway in legal theory. And it's what lawyers mean when they tell you that "the court gives law not justice." Where did it come from? Positivism.

Just as logical positivism swept a new and intolerant bivalence to power in science and philosophy in the first half of the twentieth century, *legal positivism* won the day in law. The roots lay in Thomas Hobbes's 340-year-old book *Leviathan:* Law is the will of the legislator. In this century Hans Kelsen and H. L. A. Hart mixed Hobbes's view with the logical positivists' love of precise statements and even their symbolic logic. They saw law as a logic tree of rules or edicts. The tree starts with a *Grundnorm* or grounding norm or value statement such as you might find in a key phrase of a nation's constitution. Then a law or legal rule is valid if, and only if, a lawmaker says so or a tree branch connects it to an earlier rule that a lawmaker says is valid. Validity flows up the tree from the root.

So why not replace a judge with a computer? Law is a rule tree, the positivists claim. A judge just checks the rules and matches the facts to rules. She may get that wrong or play out her prejudices or give hell on her bad days. A computer plays it straight by the rule book. Law students hear about this computer judge when they take a class in jurisprudence. The teacher sets it up so he can chop it down with principles.

The new view in law is that rules sit in a nest of principles. The rules come and go as the times change. The principles change slowly as the times change. Each case invokes all principles to some degree. Fresh cases stretch and flesh out the old principles. But most rules die young. As Roscoe Pound put it,

> The vital, the enduring part of the law is in principles—starting points for reasoning, not rules. Principles remain relatively constant or develop along constant lines. Rules have relatively short lives. They do not develop; they are repealed and are superseded by other rules.*

Ronald Dworkin has built a whole theory of law on the split between rules and principles. Principles, Dworkin says, "have a dimension that rules do not—the dimension of weight or importance," and the court "cites principles as its justification for adopting and applying a new rule." We have volumes full of rules but only a few of principles. Rules allow or disallow precise acts. Principles guide us when we pick and kill rules:

---

*Pound, R., "Why Law Day," *Harvard Law School Bulletin*, vol. X, no. 3, 1958.

> Only rules dictate results, come what may. When a contrary result has been reached, the rule has been abandoned or changed. Principles do not work that way; they incline a decision one way, though not conclusively, and they survive intact when they do not prevail.*

On the principles view a judge must first weight all principles and then add them up to decide the case. All the principles "hang together" as Dworkin says. The judge may work with some clear rules but this is rare. The hard part is matching fuzzy facts to all "relevant" principles. You contest a verbal contract with your boss or car dealer or landlord. The case comes to court. The judge matches your facts to the principles of freedom of contract, full disclosure, buyer beware, seller beware, and acquiescence. You were free to large degree when you struck the deal. You were to large degree open and clear on the terms. You acted on it to some degree. The judge weights the principles and cites case precedents to back up the weights. The judge does not give the weights as numbers—at least today a judge does not—but they are a matter of degree. Some weights rank more important than others. Some judges know more law and see more connections and cite more cases than others do. The judge cites these cases to justify her ruling. She does not point out an audit trail in a rule book. She gives what looks a lot like a fuzzy weighted average.

The FAM analogy to a law judge is only an analogy. But the FAM model comes closer to how we judge things than the decision tree model comes. It gives a new way to think about how we reason. And law principles look a lot like fuzzy rules.

We now drop the analogy and turn to where we have built systems in the FAM image.

FAMS IN PRACTICE:
FUZZY PRODUCTS

How does a fuzzy washing machine work? At the math level it acts as a FAM. Rules pick out patches and the patches add up to cover the system. Most new machines have a small chip or

---

*Dworkin, R. M., *Taking Rights Seriously*, Harvard University Press, Cambridge, MA, 1977.

microprocessor in them. You program the chip to store the FAM rules and make decisions as fuzzy weighted averages.

At the user level you drop clothes in the washer and press Start. In some models once a week or so you also load the machine with detergent. The smart washer adjusts the wash cycle to the type of clothes and their dirt status. Some stains take longer to dissolve than do others. Oil stains break down slowly. Mud and dirt break down quickly. The agitator acts as a load sensor as it turns the clothes. The agitator feels the load size and guesses at the clothing type as it knocks off dirt. It sends this data to the microprocessor. As the machine continues to wash, an optical sensor pulses a light beam through the pipe of waste water. The wash murk clouds the light pulse and tells the sensor, which tells the microprocessor, the dirt level, and the detergent level. Some new machines shoot bubbles into the wash water to help break up dirt and detergent. All fuzzy washing machines help prevent cloth damage and underwashing and overwashing.

The fuzzy system sits in the microprocessor. That is where fuzzy systems sit in all smart products. The fuzzy system turns sensor data into wash commands. There are 600 or so commands it can give. Fuzzy weighted averages pick the commands. Each second or each fraction of a second the FAM system tells the agitator how to turn and whether to let in more or less water and how to change the wash or spin or rinse cycles or whether to repeat them. The engineers at Hitachi and Matsushita (Panasonic) and South Korea's Samsung drew the fuzzy-set triangles for the load size and water clarity and water flow and then related them with 30 or so rules. The load is small, medium, large, or very large. The water is very dirty, dirty, medium, clean, or very clean. Let in no water, a little, medium, or a lot of water. If the load is heavy and dirty, use a lot of water and repeat the cycles. If the load is light and clean, use a little water and do not repeat a cycle.

In the same way, a fuzzy dryer turns the flow of hot air, load size, and fabric type into drying times and drying strategies. Fuzzy microwave ovens measure temperature, humidity, infrared light patterns, and change in food shape and then map these to cooking times and patterns of hot-air blowing.

New sensors help the washing machine and help all smart machines. Critics point to the new sensors when they want to deny that fuzzy logic itself raises machine IQs. New sensors help and no one disputes that. They give more data and they give it faster and more precisely than old sensors or no sensors give it. But competitive markets are efficient, as economists say. The Japanese and Koreans and Taiwanese do not use fuzzy logic because it fits well with Zen or their personal world view or because it gives them a new buzz word to help sell goods. That is all after the fact. More money fits best with most world views. The November 1991 *Fuzzy Logic* report of the U.S. Commerce Department calls out this new blend of Zen and money:

> From a philosophical viewpoint, the fuzzy logic concept is attuned to the fundamental teachings of Zen Buddhism, which perhaps contributed to the Japanese acceptance of this concept. There are others who believe that the fuzzy logic success in Japan is a result of that country's perceived need to become competitive in advanced technologies such as artificial intelligence, biotechnology, and optical computing.

In most cases Japanese, Korean, and Taiwanese firms have tried many types of control systems tuned in many ways. They have compared each system with the others. And each system uses all the data the new sensors give it. In most machines that use fuzzy logic the fuzzy system has beaten its nonfuzzy competitors. That's why it's there—performance.

Fuzzy cameras and camcorders are a case in point. The Canon hand-held H800 cameras use only 13 rules to tune the autofocus on the lens. The fuzzy system takes up a tiny bit of memory, only 1.1 kilobytes,* a small part of even an old big floppy disk. Small electrical sensors measure the image clarity and the change in image clarity in six parts of the image. That gives 12 types of sensor data. The 13 rules turn these 12 data into new lens settings. Canon tests its cameras on a large set of images. Canon claims the fuzzy system focuses twice as well as did the other controllers they tested and used to use.

Fuzzy camcorders work much as fuzzy cameras work. Sanyo Fisher's 8mm fuzzy video camcorder, the FVC-880, uses only nine fuzzy rules. The lens image divides into six regions. The

---

*1 byte = 8 bits of information. 1 bit of information answers 1 binary question with a yes or a no. 1.1 kilobytes = 1,100 bytes. 1 kilobyte answers 8,800 binary questions.

center region counts most and the border regions count least. Old camcorders weight and add the image intensity in the six regions. The fuzzy rules tune the autofocus with the relative contrast between the six regions. Some regions are MUCH BRIGHTER or LITTLE BRIGHTER than others. Some are ALMOST EQUAL. The system first reasons with all these rules and then computes its weighted fuzzy output.

Matsushita (Panasonic) has added rules to stop the image jitter that starts when your hands shake or move. The jitter problem gets worse as camcorders shrink in size. A few fuzzy rules squelch small shakes. They guess where the image will shift and then compensate. Big palm shakes still shake the image. You can also cancel jitter with math models but they take up a lot more chip space and tend to cancel only one or two types of jitter. The engineer has to figure out the jitter type in advance and then find a math model that stops it.

The old math school works that way. Someone has to paint the problem in math terms. That works best in textbooks and classrooms. Real problems are unique and sloppy and hard to pin down in exact math terms. No one knows enough math. In some cases you can come close and get by with a bad math fit and in many of those cases you can use standard control theory to control the system. A fuzzy system may work as well or better and it uses little computer memory and you can change and polish it a rule at a time. You also don't need to know much math or control theory to build and tune a fuzzy system. Novices do it all the time and that never sits well with control engineers who have spent years studying control theory. No one likes to get beaten at her own game with someone else's rules, especially with fuzzy rules.

A fuzzy system may do nothing more for you than save chip space and make it easier to change the system. Sugeno's fuzzy helicopter is an extreme case. It breaks new ground and controls a system no man or math has yet controlled. Fuzzy washing machines make an old job a little easier. That kind of advance is small but enough to make billions of dollars when spread out over hundreds of products. Noboru Wakami, the Matsushita engineer who oversaw the first fuzzy washing machine, puts it in terms of food: "Fuzzy theory is like seasoning. Sometimes the seasoning simply improves the taste. Sometimes it produces something dramatically different."

A fuzzy vacuum sweeper only improves the taste. As always the fuzzy system sits in a small chip. Infra-red sensors detect the carpet or floor type and measure the amount of sucked dirt. The dirt and floor data fire the fuzzy rules. If the floor is shag carpet, suck hard. If the floor is tile, suck little. If the carpet is very dirty, suck very hard. Out comes a weighted fuzzy average that gives the suction motor's power in watts. The fuzzy control is elegant and saves watts but here that may not be too big a deal. It may save only a few cents per room. Fuzzy air conditioners save still more watts. Mitsubishi and Korea's Samsung report 40% to 100% energy saving with their fuzzy models.

Fuzzy car systems do more than improve the taste. The major Japanese and Korean car firms have built and patented many car systems. Nissan holds patents on fuzzy systems for antiskid brakes that pump the brakes in some optimal way and on automatic transmissions that shift gears based on road and car conditions as the best human drivers do with a stick. The firms will not reveal the exact type or number of fuzzy rules. Nissan uses a set of rules to control fuel injection in engine cylinders. Sensors measure the manifold pressure, throttle setting, water temperature, and r.p.m., and feed this data to a small onboard microprocessor. There a FAM system converts it to a fuel flow. A second FAM system combines data on r.p.m., water temperature, and oxygen concentration to time the ignition. Mitsubishi uses fuzzy systems to control suspension, air conditioning, transmission, and four-wheel drive.

Here is a 1992 list of fielded fuzzy products in Japan and South Korea:

Product	Company	Fuzzy Logic Role
Air Conditioner	Hitachi, Matsushita, Mitsubishi, Sharp	Prevents overshoot-undershoot temperature oscillation and consumes less on-off power
Anti-lock brakes	Nissan	Controls brakes in hazardous cases based on car speed and acceleration and on wheel speed and acceleration

Product	Company	Fuzzy Logic Role
Auto engine	NOK/ Nissan	Controls fuel injection and ignition based on throttle setting, oxygen content, cooling water temperature, RPM, fuel volume, crank angle, knocking, and manifold pressure
Auto transimssion	Honda, Nissan, Subaru	Selects gear ratio based on engine load, driving style, and road conditions
Chemical mixer	Fuji Electric	Mixes chemicals based on plant conditions
Copy machine	Canon	Adjusts drum voltage based on picture density, temperature, and humidity
Cruise Control	Isuzu, Nissan, Mitsubishi	Adjusts throttle setting to set speed based on car speed and acceleration
Dishwasher	Matsushita	Adjusts cleaning cycle and rinse and wash strategies based on the number of dishes and on the type and amount of food encrusted on the dishes
Dryer	Matsushita	Converts load size, fabric type, and flow of hot air to drying times and strategies
Elevator control	Fujitec, Mitsubishi Electric, Toshiba	Reduces waiting time based on passenger traffic
Factory control	Omron	Schedules tasks and assembly line strategies
Golf diagnostic system	Maruman Golf	Selects golf club based on golfer's physique and swing

Product	Company	Fuzzy Logic Role
Health management system	Omron	Over 500 fuzzy rules track and evaluate an employee's health and fitness
Humidifier	Casio	Adjusts moisture content to room conditions
Iron mill control	Nippon Steel	Mixes inputs and sets temperatures and times
Kiln control	Mitsubishi Chemical	Mixes cement
Microwave oven	Hitachi, Sanyo, Sharp, Toshiba	Sets and tunes power and cooking strategy
Palmtop computer	Sony	Recognizes handwritten Kanji characters
Plasma etching	Mitsubishi Electric	Sets etch time and stategy
Refrigerator	Sharp	Sets defrosting times and cooling times based on usage. A neural network learns the user's usage habits and tunes the fuzzy rules accordingly.
Rice cooker	Matsushita, Sanyo	Sets cooking time and method based on steam, temperature, and rice volume
Shower system	Matsushita (Panasonic)	Suppresses variations in water temperature
Still camera	Canon, Minolta	Finds subject anywhere in frame, adjusts autofocus
Stock trading	Yamaichi	Manages portfolio of Japanese stocks based on macroeconomic and microeconomic data
Television	Goldstar (Korea),	Adjusts screen color and texture for each frame and stabilizes

PRODUCT	COMPANY	FUZZY LOGIC ROLE
	Hitachi, Samsung (Korea), Sony	volume based on viewer's room location
Translator	Epson	Recognizes, translates words in pencil-size unit
Toaster	Sony	Sets toasting time and heat strategy for each bread type
Vacuum cleaner	Hitachi, Matsushita, Toshiba	Sets motor-suction strategy based on dust quantity and floor type
Video camcorder	Canon, Sanyo	Adjusts autofocus and lighting
Video camcorder	Matsushita (Panasonic)	Cancels handheld jittering and adjusts autofocus
Washing machine	Daewoo (Korea), Goldstar (Korea), Hitachi, Matsushita, Samsung (Korea), Sanyo, Sharp	Adjusts washing strategy based on sensed dirt level, fabric type, load size, and water level. Some models use neural networks to tune rules to user's tastes.

There are many more applications. And behind them all is a legal thicket of patents.

How do you patent a fuzzy system? They all use the same FAM architecture or some small variation of it. You can't patent that because you can't patent math. I did not patent the FAT theorem and could not have if I had wanted to. Do you patent the fuzzy rules? Some firms try to do that. But most systems use the same kind of rules. Most control systems are "error nulling" systems that turn left when something turns right or that go down when something goes up. They try to reduce or null the error or gap between where the system is and where you want it to go. That leads to similar rules and most of these rules are just a short list of software. You may be able to patent the box that houses and implements a fuzzy system. This breaks down for fuzzy chips since they use the same logic components. Then it reduces to chip design or software instructions and again these tend to be very similar. Lawyers have asked me to play expert witness in Japanese patent disputes. That honor I decline.

Japanese firms hold over a thousand fuzzy patents in Japan. As of December 1990 they also held 30 of the 38 fuzzy patents, many sweeping in scope, that the U.S. government had given out*:

Japanese Firm	U.S. Patent Description
1. Fuji Photo Film	Liquid and powder measuring device
2. Fuji Photo Film	Powder measuring device
3. Fuji Photo Film	Method of measuring liquid
4. Fuji Photo Film	Control method and measuring method for liquids and powders
5. Mitsubishi	Power system stabilizer
6. Mitsubishi	Auto-tuning controller
7. Hitachi	Fuel-injection controller for internal combustion engine
8. Hitachi	Device for stopping vehicle at predetermined position
9. Hitachi	Analogical inference method and apparatus control system
10. Hitachi	PID controller
11. Omron Electronics	Fuzzy data communications system
12. Omron Electronics	Fuzzy semifinished integrated circuit
13. Omron Electronics	Fuzzy function circuit
14. Omron Electronics	Fuzzy logic computers and circuits

*Source: U.S. Patent and Trademark Office, U.S. Department of Commerce.

JAPANESE FIRM	U.S. PATENT DESCRIPTION
15. Omron Electronics	Fuzzy logic basic circuit and integrated circuit operable in current mode
16. Omron Electronics	Fuzzy membership function circuit
17. Honda	Vehicle control system A
18. Honda	Vehicle control system B
19. Japan Electronic Control Systems	Electric air-fuel ratio controller
20. Japan Electronic Control Systems	Air-fuel mixture ratio controller for internal-combustion engine
21. Japan Electronic Control Systems	Electronic learning control apparatus for internal-combustion engine
22. Toshiba	Adaptive process controller
23. Toshiba	Apparatus for performing group control on elevators
24. Toshiba	Automatic trouble analyzer
25. Matsushita Electric	Temperature-adjustable water-supply system
26. Mazda	Control system for vehicle engines
27. Nissan	Fuzzy control system for automatic transmission
28. Nissan	Vehicle air-conditioning system based on fuzzy inference
29. Nissan	Antiskid braking control system based on fuzzy inference
30. Agency of Industrial Science and Technology	Method and apparatus for recognizing colored patterns

Americans have heard that the Japanese think business is war. Pop books and movies and talk shows have made sure they hear it. It scares them and makes them mad and sometimes makes them call for new trade laws. A glance at that patent list might make some call a little louder. The extreme view sees the Japanese in their 1940s war mode but with corporate buyouts and land grabs and fists full of patents in place of guns and planes and war ships.

That view forgets that business is still business even if it is war. From the outside Japan may look as if it acts with one purpose. But inside the hive it is every firm for itself.

I want now to describe the two large fuzzy centers in Japan that are at the heart of fuzzy systems in the fuzzy present. I have toured and lectured in both centers and support both and know the people who run them and know many who work in them. The centers bring together Japanese business, government, and culture. From the outside they look almost like war machines of the Information Age. Inside and up close they just look like the chaos of war.

LIFE IN JAPAN

The Japanese government has helped pay for two fuzzy research centers in Japan. On March 28, 1989, the Japanese Ministry of International Trade and Industry (MITI) launched the Laboratory for International Fuzzy Engineering Research (LIFE) in Yokohama, just south of Tokyo. LIFE gets $70 million over five years. It will no doubt live another five years after the first five are up.

LIFE has a rival. On March 15, 1990, MITI helped launch the Fuzzy Logic Systems Institute (FLSI) in the south of Japan, at the Kyushu Institute of Technology on Kyushu Island. The top Japanese firms in electronics, automobiles, and manufacturing belong to LIFE or FLSI or both.

A bureaucracy runs LIFE. One man runs FLSI: Dr. Takeshi Yamakawa. The two centers cooperate and compete. But they mainly compete and that is the true but unsung big picture of fuzzy logic in Japan.

Dr. Toshiro Terano, of the Tokyo Institute of Technology,

officially heads the LIFE project. But young Dr. Tomohiro Takagi, on loan from Panasonic, runs the lab. Both are smart and polite men and have worked in fuzzy logic. But these two men did not put LIFE together.

The real force behind LIFE is Terano's former Ph.D. student Professor Michio Sugeno of the Tokyo Institute of Technology. Sugeno built the fuzzy helicopter and is, as I have said, the finest fuzzy math theorist in Japan. In the early 1970s he developed in his Ph.D. dissertation a fuzzy integral (as did I in my dissertation) that gives the fuzzy version of the dreaded calculus that young scientists and engineers must learn in college or high school. Sugeno is aloof, parts his hair in the middle, and can do without Westerners. He was five in 1945 when the Allies bombed his city of Yokohama.

You can get a feel for LIFE's business might when you look at its board of directors. Out front stands Katsushige Mita, one of Japan's gray-haired captains of industry. Mita is the president of Hitachi and chair of LIFE's board of directors. The board roster says it all and has no counterpart in Europe or the United States.

LIFE BOARD OF DIRECTORS

Chairman	Katsushige Mita	President, Hitachi Ltd.
Executive Director	Masato Nakayashiki	
Managing Director	Dr. Toshiro Terano	Professor, Hosei University
Director	Joichi Aoi	President, Toshiba
Director	Tetsuro Kawakami	President, Sumitomo Electronics
Director	Yutaka Kume	President, Nissan Motors
Director	Yutaka Saito	President, Nippon Steel
Director	Fusaro Sekiguchi	President, Meitec
Director	Hideo Tajima	President, Minolta
Director	Akio Tanii	President, Matsushita Electronics

Director	Yoshio Tateishi	President, Omron Electronics
Director	Kei Yamaguchi	President, Ebara Research
Director	Takuma Yamamoto	President, Fujitsu

At LIFE's inception in March 1989, 48 Japanese companies paid varying fees to join LIFE as associative members:

Canon
Ebara Research
Fuji Electric
Fuji Heavy Industries
Fujitsu
Fuji Xerox
Fuyo Data Processing and Systems Development
Hitachi
Honda
IBM Japan
INTEC
Japan Aviation Electronics Industry
Japan Electronic Computer
Kao
Kawasaki Steel
Kayaba Industry
Kobe Steel
Konica
Kozo Keikaku Engineering
Matsushita Electric Industrial
Mazda Motors
Meidensha
Meitec
Minolta
Mitsubishi Electric
Mitsubishi Kasei
NEC
Nippon Steel
Nissan Motors
NKK
Nippon Telephone and Telegraph (NTT)

Oki Electric Industry
Olympus Optical
Omron Tateisi Electronics
Sharp
Shimizu
Sony
Sumitomo Electric Industries
Sumitomo Metal Industries
Takenaka
Thomson Japan
Tokyo Electric Power
Toshiba
Toyota Motors
Yamaichi Securities
Yamatake-Honeywell
Yazaki
Yokogawa Electric

Katsushige Mita has stated his view on fuzzy logic and the role it and LIFE will play in the Information Age:

> Information technology has made remarkable progress. Intensive research is now aimed at the realization of artificial intelligence. However, operations of present computers depend on simple Yes/No logic, namely binary logic, which is widely different from the information processing inherent in human thinking. Therefore, evaluation based on common sense and flexible judgment is considered difficult to achieve by computers.
>
> Fuzzy theory has emerged as a theory suited to represent uncertainty contained in the meaning of each word. Fuzzy artificial intelligence, as an application of this theory, is expected to play an important role in the future establishment of an intimate relationship between men and computers. It is indispensable for configuring human-friendly machines and speech-recognizable systems such as nursing and home robots, and for developing AI tools which support man in production control, medical diagnosis, finance and securities, general decision making, etc. From this perspective, we have founded a technological research association, "Laboratory for International Fuzzy Engineering Research (LIFE)." This organization is intended to vitalize basic study of fuzzy theory, research on its efficient utilization by strengthening ties between industrial and academic circles, and

> to promote international technological exchange. All of these activities are aimed at promoting the early advent of a highly sophisticated information-intensive society.
>
> Our laboratory will always view information technology as an adjunct to man, respect the subjectivity and personality of each individual in the information-intensive society, and strive to create technology for mankind in its true sense.

LIFE has many goals and holds a conference every other year in November in Yokohama. LIFE held its first conference in 1991. The rival center FLSI in Iizuka City on Kyushu Island in the south holds an international conference on fuzzy logic and neural nets in July in even years. I was program chair of the first meeting, Iizuka-90, and program co-chair of the second meeting, Iizuka-92. So each year Japan holds a major fuzzy conference. The U.S. held its first major fuzzy conference in March 1992 in San Diego. Only 500 people showed up. That cut short the scenes a documentary crew had planned to film there. And almost no Japanese came and only one Japanese company had an exhibit booth. It was a quiet Japanese boycott.

LIFE did have three large laboratories that ran twelve or so projects. The first lab was for fuzzy control. The second lab, which Dr. Tomohiro Takagi headed, looked at "fuzzy intellectual information processing." This lab had the "soft" slant of computer science and worked with fuzzy databases, pattern recognition, decision systems, natural language processing, and cognitive processing. It was the first to apply the fuzzy cognitive map (discussed in the next chapter) to real problems of plant control and scheduling. The third lab worked on fuzzy computers and chips.

In 1991 LIFE reorganized. The three large labs dissolved into their twelve or so separate projects and they all went under Dr. Takagi, who goes by Tom. I had lectured at LIFE in the summer of 1989 and saw the reorganization coming. With so many men from so many companies that compete, and with each man's loyalty to his company so fierce, the politics evolve fast.

I met then in private with the LIFE leaders and many of LIFE's supporters. They all spoke softly and complained of the lack of concrete goals. Their own companies worked with goals and subgoals and subsubgoals right down to the limits of micromanagement. There was little trust or real sharing. They worried that they worked too closely with their competitors and felt some researchers got a free ride and pots of money while others, like them, had to fight to keep what little money they had. They suspected that some companies kept their best researchers at home and sent LIFE their poorer workers and their best spies. Yet the free form of research excited them. They were proud to lead the world in fuzzy machine intelligence and to finally "do research" as the Americans do. This meant even middle-aged Japanese men could act a little like *shinjinrui*, the new breed of unorthodox Japanese youth who vote for leisure over work and watch MTV and sometimes dress like their rock-star idols.

The Japanese I know have heard all the slights about how the Americans invent things and the Japanese only improve what Americans invent. I hear this from them every time I go to Japan. It tends to come up when we drink large bottles of Japanese dry beer.

I remember a dinner I had once in Kyoto with some Omron bosses, one of the main LIFE and FLSI sponsors. At the time an Omron manager ran one of the three LIFE labs, so the Omron bosses knew all the latest LIFE politics. Omron wants very much to be first in fuzzy logic in Japan and the world. We went to a fine Japanese restaurant where the women wear kimonos and bow deeply and say "*arigato gozai mas*" over and over. The bosses gave me the "first" seat, facing the door. You do everything in Japan by rank. The guest ranks first unless the big boss joins you. The talk ran out fast since their English was bad and my Japanese was far worse. I had read speed-Japanese books on the flight over and watched as many Akira Kurosawa movies as I could. Drink filled in for talk. In Japan you drink beer from small clear glasses and you cannot let a guest's glass stay empty. So soon I had drunk several bottles of dry beer and so had my Omron friends. We munched whole pickled bluefish and strange black and green spicy tofu cubes. No one knew what was in the tofu cubes besides salt. We just washed them down with more dry beer.

The talk had so stalled that my friends now talked to each other in Japanese. They knew this was rude and soon stopped and poured more dry beer and tried to make small talk. That dry beer was all we seemed to have in common. So we talked about dry beer (it has less unfermented sugar in it). They asked me if I liked it. I said I did and emptied my glass. They filled it and asked me if I drank dry beer in Los Angeles. One of the four men had been to California and the other three wanted to go. I said yes and told them how glad I was that you could now buy Japanese dry beer in California stores. They nodded and said nothing. I searched for what to say next to keep the talk going. I said I wished Budweiser well in their new dry beer. This took two or three tries to get across. They were shocked that there was a U.S. dry beer. It was fun to shock them. To shock them some more I told them that more U.S. dry beers were coming and that the Europeans had them now too. The lead Omron boss picked up a big half-full bottle of Asahi Super Dry and set it on the palm of his right hand. He turned to his friends and the drink made him forget to speak in Japanese. He said in English, "The Americans have stolen another idea from us!" We all laughed.

The LIFE folks talk about more than dry beer when they eat. These can be the men who work in LIFE or the many more who work with fuzzy logic in their companies that support LIFE. Women work in fuzzy logic too in Japan but they never come to dinner. The men talk of many things but the ones I talked to all agreed on this: Takeshi Yamakawa has got to go.

After a few bottles of dry beer and some sake the talk always comes round to Yamakawa and his new Fuzzy Logic Systems Institute in the south. They ask me if I support him. I tell them yes and that I admire him. They agree. They support him and admire him too. Then it starts. Yamakawa's research is derivative. His analog fuzzy chips don't work. Or the chips are too big and bulky and cost too much and are out of date. Yamakawa acts too much like an American. No offense. He made too much money on his fuzzy book *The Concept of a Fuzzy Computer*. His wife really wrote it. It had no meat in it. It went to his head. He's deceptive and too ambitious. Sure we admire him. But he acts too much like an American. He oversells. He started his own FLSI journal. You see him on TV. He deals with your NASA as if he speaks for Japan. You don't understand. You can't act that way in Japan. Japan is different. You don't ignore the order and set up your own shop. You don't start a new company or new center in Japan. You work within the system. Kyushu is too hot and sticky a place to hold a conference in July. Westerners hate it there. Half the toilets are the old pits in the ground. Yamakawa has no contract support. His contracts won't last. He may be a hero on Kyushu Island but not here. He may know karate but research is not karate. We don't need another fuzzy research center. No one will support it. Who does he think he is? No one heard of him until a few years ago. Why won't he work with us instead? He won't come to LIFE. He acts too much like an American.

Yamakawa says one thing to all that: "Life is short."

YAMAKAWA IN THE SOUTH

Yamakawa means mountain river. Professor Takeshi Yamakawa likes that image of water flowing down a big mountain, adapting to it, wearing it down. Yamakawa is short and in his forties and has gray frost in his thick black hair. In the spring of 1992 he earned his *fifth* dan black belt in Shotokan karate. This is the real thing and not like the new karate you find in the U.S. in small one-room karate gyms in small shopping malls. Yamakawa took up karate after he got smashed up in a car wreck. Since the wreck he rides a bike or walks or takes a taxi or train but never drives a car.

I remember Yamakawa at the big Tokyo conference on artificial intelligence in 1989, AI-89. Fifty thousand people went to it. More go each year. In the U.S. the big AI conferences draw no more than 5,000. Yamakawa and I gave plenary talks to the section on fuzzy logic. We drank green tea in a small room where we waited for our call to go on stage. We both knew the audience was full of his enemies who saw him as the too-American *murahachibu* or outcast. Men sat in the audience who controlled his funding and who advised those who controlled it. They might cut his contracts or denounce him to the press or write to the Kyushu Island politicians who supported him. Others just hated him. You always get some stagefright when you have to wait like this. But Yamakawa had his career on the line at the biggest AI conference in the world. He still talked and joked. Then a staffer ran into our room in a panic and said, "Time to go! All fill up! Must go!" I felt the small butterflies in my solar plexus. Yamakawa and I stood up and picked up our stacks of view graphs. I looked at him and felt sorry for this man all alone in the midst of his enemies. Still I like a good fight and wanted to see if he could handle the pressure. I gargled some warm water to loosen my throat and then grinned and asked

Yamakawa if he was ready. He said nothing. He passed his view graphs from his right hand to his left hand and then dropped into a deep front stance as a fencer might. Then he threw a perfect right front punch in the air. It made a popping sound as he shouted a muffled "*Kiai!*" The old *bushido* code. That was why his talk went well and why sometimes one man can beat many.

Yamakawa runs the Fuzzy Logic Systems Institute (FLSI) through the new Kyushu Institute of Technology. KIT wants to be the CalTech of south Japan. KIT is but one of over twenty new high-tech centers that the Japanese government has started in its plan for a "technopolis" in the twenty-first century. The Japanese know that knowledge, not just information, is the commodity of the future.

Today you do not find much of the old Japan in Japan. The old traditions have given way to TV and a few festivals. What you find, outside of the shrines and temples in Kyoto, tends to be in the south on Kyushu Island. Yamakawa is as traditional as his northern Tokyo colleagues are modern. He starts every day with a half-hour of *zazen* meditation. He just sits in full-lotus position. His whole family meditates. He is an expert at the Japanese bamboo flute, the *shakuhachi*, and his wife is an expert at the plucked three-stringed *samisen*. They give joint recitals and wear traditional flowing robes. They met when Yamakawa came to her for music lessons.

Yamakawa does a lot more than sit still for a half hour each day. He teaches engineering at Kyushu Institute of Technology and guides students in new research and runs the Fuzzy Logic Systems Institute and edits its journal and does all the things one man must do to keep a center and a dream afloat and funded and popular. To do this he keeps his sleep down to four hours or so a night and carries with him at all times a portable telephone. Yamakawa also does his own new research in fuzzy systems. He builds fuzzy chips and designs new fuzzy math models and the neural fuzzy math models we will discuss in the next chapter. He also does something I have never seen done before. He builds the machines that he uses to build his fuzzy chips. He fills two rooms in KIT with chip gadgets and oscilloscopes and etching devices. And he keeps up with his two sons and his flute practice and his karate. This mix of high-tech and Zen tradition lies at the heart of the man who stands for fuzzy systems in Japan.

Takeshi Yamakawa grew up in the green forested hills of nearby Kumamoto City. He took me there in May of 1990 and I used the time to stay away from phones and faxes and work on my textbook on neural nets and fuzzy logic. Yamakawa wore a gray robe and wooden shoes that clacked and no one gawked at him. The main sight in Kumamoto City is Kumamoto Castle, the third-largest castle in Japan. It had been home to the greatest hero of Japan, the samurai warrior Miyamoto Musashi, who founded kendo and who sat at the end of his life in April and May of 1645 in the cool secluded cave Reigendo and wrote the strategy book called *The Book of Five Rings* (*Go Rin No Sho*). You can now buy a pop version of the book slanted for U.S. businessmen. The legend says Musashi killed over a thousand men in combat and in sword duels. On his way to his "twofold way of pen and sword" he mastered Zen painting, poetry, and sword design. You find the Musashi story in many Japanese films and books and cartoons.

Yamakawa told me Musashi was his hero when he was growing up. It was clear to me that Musashi was still his hero. Yamakawa loathed the decadent *shinjinrui* wannabes in the North who he felt had seen too many American movies and TV shows and listened to too much pop music. He never whined about them or how they ganged up on him and tried to ruin him. When that subject came up he just waved his hand and grunted in disgust. He might say "Life is short" and laugh but that was all. To him they were soft men, the kind the world breeds most of today. They were dandies in Western suits who hid in academia and in middle management and who were part of the decline of Japan.

Yamakawa took me to Kumamoto Castle to see some of Musashi's bones. Musashi had killed so many men in sword fights that he knew some of their kin would try to foul his grave. So to hedge the risk he had his friends split up his remains and bury him in pieces in different places. Yamakawa knew all the grave sites by heart and told the taxi driver how to find them. The better graves were hidden deep in green forests of bamboo and maple and surrounded with large stones with strange meanings and with small empty Shinto shrines. Yamakawa told me what the stones and shrines meant. I have never seen him so solemn. This was his idea of what is holy. Yamakawa had no notion of Yahweh or Jesus or Allah as we have in the West. We may not believe in those notions but we have a place for them in our word schemes and mind schemes. They were no more to him than Zeus and King Arthur and Aladdin are to us. His Zen was godless. It was the Zen of the old samurai warriors who meditated to perfect their breathing and most of all their sword stroke. From what he told me at other times I knew he felt his colleagues in the North had no Zen and were just godless. They played for empty fame and gain like all the rest. They did not play to perfect something. He played for fame and gain too. But these were small ends along the way. Fame and gain helped keep score. They were footnotes to being the best you could be at something in both mind and action. That was where he found meaning in a life that may have none. He turned to Musashi in perfecting his sword fighting and poetry. That was his role model or his Jesus or his pop star.

We ended the tour with a trip to Musashi's cave back in the humid green hills outside of Kumamoto City. You hike back to the cave on steep rocky paths and through cedar trees and rock terraces covered with little stone buddhas with chipped heads. The cave opens wide and runs up into the small mountain and stops short. It was so humid that it started to rain. In Musashi's cave we at once hit cool damp air. Yamakawa and I each sat on a rock and watched the rain fall in straight lines outside the mouth of the cave and in layered sheets across the valley on the orange groves and green cedar forests cut into the steep hills. I could see how you could write a book in this cool den. Yamakawa would not talk. He was in Mecca. We just sat there for a half hour until the rain stopped. *Zazen*. Just sitting.

11

ADAPTIVE FUZZY SYSTEMS

From causes which appear similar, we expect similar effects. This is the sum total of all our experimental conclusions.

DAVID HUME
AN INQUIRY CONCERNING HUMAN UNDERSTANDING

A learning machine is any device whose actions are influenced by past experiences.

NILS NILSSON
LEARNING MACHINES

The brain is a universal measurement device acting on the quantum level.

STEPHEN GROSSBERG
STUDIES OF MIND AND BRAIN

Brains are computers made of meat. Our meat computers are dumb or smart. It depends on the brain's hardware of ''wetware'' and on the brain's software or training and experience.

Fuzzy systems can be dumb or smart. It depends on the fuzzy rules. Some engineers come up with very clever fuzzy rules. Some just guess and get the rules wrong. Some work at it and still get them wrong. Fuzzy cameras and washing machines and subway engines use only a few fuzzy rules. Engineers tested and tuned them for months. The subway in the city of Sendai uses 54 rules. Hitachi engineers honed and tested these rules for years. Now with software you can build a fuzzy system in minutes and then spend hours or days trying out new rules and redrawing their fuzzy sets to tune the rules. Or you can use a neural system to tune them for you.

It all comes down to a search for fuzzy rules. So far that search has depended on the human brain. The goal is to take our brains out of the loop and give fuzzy systems their own brains, their own way to grow their rules. This chapter looks at this new frontier of *adaptive* or learning fuzzy systems. We start with finding rules in data clouds and end with a new tool to help predict the rise and fall of governments.

THE DIRO BRAIN SUCK:
DATA IN, RULES OUT

The FAT theorem tells us that *in theory* we can always find fuzzy rules to simulate or approximate any type of control or computer processing. But *in practice* we may have no idea where to begin.

Abstract theorems often leave us in such a fix. In the 1930s mathematician John von Neumann proved that chess and checkers and tic-tac-toe and other parlor games have optimal strategies that always win or tie. Von Neumann's theorem of "game theory" proved that these optimal strategies exist. It did not show how to find them. So far we have found the optimal strategies for tic-tac-toe and checkers. But the search goes on for optimal chess strategies and may go on for centuries.

The FAT Theorem is more helpful. It suggests ways to automate the search for good rules. As we will see, *neural networks* or brainlike computer systems can help find these fuzzy rules.

We seek a system that learns fuzzy rules from experience. That means it learns from data. Today we ask experts how to focus a camera or land an airplane or shoot down a SCUD missile. Fuzzy engineers turn expert talk into fuzzy rules. They use a ghost that hides in their brains, a ghost called *judgment*.

We seek a learning system that turns expert *behavior* into fuzzy rules. The experts leave footprints in the data. They leave a number trail that the adaptive system converts into fuzzy rules. The rules leave a similar number trail, similar footprints in the data. The adaptive system sucks the brain of the expert or the computer or whatever came up with the data. The more data, the better the brain suck, the finer the fuzzy rules. Data to rules.

DIRO: Data in, rules out. Numbers in, knowledge out. Experience in, common sense out. Examples in, expertise out. A neural network can fill the DIRO black box (Figure 11.1).

FIGURE 11.1

A neural net acts like the eyes and ears of an adaptive fuzzy system, a fuzzy system whose rules change with experience. The neural net senses fuzzy patterns in the data and learns to associate the patterns. The associations are rules: If fuzzy set A, then fuzzy set B.

The fuzzy system acts at a higher cognitive level to reason with the fuzzy rules. It infers or decides an outcome action based on the incoming data or facts. An adaptive or neural fuzzy system changes or tunes its rules as it samples new data. Just as every pattern we see or hear or taste or feel changes slightly our world view—remember how your ideas of tall people or a good film or good food or good music or a lot of money changed as you passed from child to adult—so every new example of expert behavior changes slightly the rules in an adaptive fuzzy system. At first the rules change fast. This lets the fuzzy system find a rough working set of fuzzy rules. Then with more samples, more expert examples, the rules change more slowly as the fuzzy system fine-tunes its knowledge. With neural nets and with brains practice makes perfect.

Consider backing up truck and trailer to a loading dock in a parking lot. Truckers do this every day. But no one has yet worked out the math. It's too hard, too messy, too nonlinear.

Truckers learn to back up their rigs by trial and error and by feel. First they learn how to back up a car. That skill lets them back up a truck cab. But put a trailer on the cab and you have to practice. You put it in reverse and look behind you and pump the gas and the trailer goes the wrong way. So you turn the steering wheel the other way and the trailer backs up the right way. You do this until you do it right without thinking about it—until it *feels right*. Practice makes perfect by making a habit and then refining the habit. Your backing skill improves a lot at first and then improves slowly and then does not improve at all. You grow and then "max" out. First you run up the "learning curve" and then you only crawl up it.

You need no verbal skill to learn to back up the truck and trailer. You do not have to talk your way through it. You just have to practice it. I have found that most truckers cannot explain how they back up though they are glad to show you how. And I have asked a lot of them.

In 1989 my Ph.D. student Seong-Gon Kong (now a professor at Soongsil University in Seoul, Korea) and I trained a fuzzy system to back up a simulated truck and trailer. I had heard that a Stanford graduate student had taught a neural net to back up a simulated truck and trailer. I wanted to show the science world that an adaptive fuzzy system could do more. It could both learn the skill and *explain* the skill since it learned the implicit fuzzy rules that produced the skill.

The system worked. You could open the black box and read the fuzzy rules. Even truckers don't do that. The truckers are "black boxes" just as are raw neural nets. They drive and say some things about how they drive but they can't fully explain what they do. For that matter you won't find any violinists or woodcarvers or aerobics teachers who can explain how they do what they do. They can show you and train you and say things that point to how they learned to do it and how you can learn to do it too. Their words fall far short of the kind of full explanation a computer needs to do the same thing. You can no more play the violin or carve a statue or lead a dance class from just their words than you can figure out what people talked about at a party from just their footsteps on the floor.

We trained our adaptive fuzzy system with the paths that a

truck and trailer makes when it backs up to a loading dock in a flat parking lot. We did not back up a real truck but simulated one on a computer. I had worked out the neural math a year before and this let us test it. We put in a few hundred truck paths and out came 105 fuzzy rules. We fed the neural net a few hundred more truck paths. Then only a few of the rules changed. Learning had slowed or "converged." This new set of rules, the new fuzzy truck backer-upper, backed up better than the first set of rules did. The new system found smoother paths and shorter ways to back out of jackknife starts. We then looked at the rules and found that they made sense. Most things make sense in hindsight. They were correction rules: If the truck turns too much to the left, steer to the right to move the truck back to the right. The truck rules looked a lot like the rules we tried to come up with and the ones that once in a while we got a trucker to tell us.

How did the fuzzy truck learn its rules? What is a neural net? What is learning? We have to go back to the FAT theorem. We saw in the last chapter that rules are patches. So the question is how you learn patches. The answer is patches are clusters in the data. Learn data clusters and you learn patches and you learn rules and you have an adaptive fuzzy system. That's the big answer but it depends on learning. So what is learning? That answer depends on neural nets. You will hear more and more about neural nets in the next years as machines grow smarter and as we "reverse engineer" the brain, as we take the brain apart a net at a time and a molecule at a time and then rebuild it in new ways in silicon and in holographic light.

LEARNING AS CHANGE: Neural Nets in Brains

To learn is to change. And to change is to learn. You can learn well or badly. But you cannot learn without changing or change without learning.

But *what* does learning change? Your arm muscles change and your brain changes when you learn to throw a curve ball or a discus or a front punch. Your world view changes and your brain changes when you learn Latin or a new religion or the integral calculus. Your behavior changes and your brain changes when you learn that fire burns or AIDS kills or cops give tickets for speeding.

Your brain changes. Three pounds of meat changes. It changes a little bit every time you see an image or hear a sound or feel a surface or taste a flavor or walk a new ground. Everything you sense changes your brain. Your brain measures things and those things change your brain. It learns new changes and forgets or unlearns old changes. A single photon of light changes your brain. TV ads and movies and books and talks and ad jingles and all the stuff that your mind eats change your brain. It changes right now as you read this.

Your brain starts out with about 100 billion neurons or brain cells and ends up with several billion fewer. That's about as many stars as are in the Milky Way galaxy or as many galaxies as are in the known universe. The neurons do not act as computer memory sites. No cell holds a picture of your mom or the smell of lime or the idea of God. You can pull out any cell in your brain and your mind will not change. You could pull out a few million cells at random and not miss them. Pull a few wires or circuits out of a computer and it crashes.

What counts is the wires between the cells. They count for about 40% of your brain mass. We call these wires synapses or neural connections. Each neuron in your brain can connect up to 10,000 other neurons. That gives synapses in the quadrillions. Learning and memory lie in the great tangled webs of synapses. Not in cells, in webs. Take a drink of scotch and lose a few hundred neurons and a few million synapses. Learning is change. In brains that means learning is change in a synapse.

Picture a huge city where each house has 10,000 phone lines that connect to 10,000 other houses. A lot goes on in each house but in the end each house sends and receives messages over the phone lines. The catch is each house sends the *same* message to all 10,000 houses. Think of the junk phone calls. Think how fast bad news and gossip would spread. Think how it would still spread if you took out a few thousand houses. The talk net is "fault tolerant."

Now think of the phone lines as plastic pipes. They grow in thickness if you talk a lot on them and shrink if you don't. They grow as a muscle fiber grows if you stress it. So the whole talk system moves. It changes with talk. It learns the talk.

Neurons are not houses. They are more like dams that fill up with electrical charge and then burst. Each neuron sends small electrical pulses or spikes down the wires to the other neurons. Each spike is a dam burst. The more spikes, the thicker the wire. The spikes are like on-off messages that pour into the neuron from the synapses. The charge builds up and then the electrical dam bursts and the neuron sends out a spike of its own to 10,000 other neurons. This makes the other neurons send more spikes out and back to the first neuron. And round and round it goes. Every neuron is out for itself like a busy shopper in a large market economy. We call it massive feedback. And we call the whole swirling wet mess a neural network.

A neural network is a set of interconnected neurons and synapses that acts as a computer and converts inputs to outputs. These mathematical or "artificial" neural nets run on software or on special chips. Today we use them to learn and recognize abstract patterns: a safe pap smear, a high-risk loan applicant, a handwritten zip code, a bomb shape in an X-ray scan of a suitcase.

How can a biological neural net learn a pattern like a face or piece of music or the concept of God? It changes millions of synapses. Zoom in closer and that means it changes the release rates of neurotransmitters. Electrical spikes flow down the cable and right before they pour onto a neuron they hit the synaptic junction. Here the electrical signal turns into a chemical or neurotransmitter signal. The chemical diffuses across a thin moat of brain juice and onto the surface of the neuron, where it changes the electrical state of the neuron. Release rates. Replenish rates. Bags of chemicals dumped into the brain juice.

Learning changes how much neurotransmitter the synapse releases and how fast. Norepinephrine (NE) or brain adrenalin is a key transmitter. Drink a cup of coffee or a glass of iced tea and for a few minutes your IQ goes up. You are more alert. If you have to learn something new or take a test or give a speech at a meeting, this is the time to do it. Your brain synapses burn up more NE. They do not make more. They just burn it up faster. Cocaine burns it even faster. (Mescalin in peyote cactus acts like NE and burns things and in ways we do not know.) But what goes up can come down even harder. You lose those IQ points and then some and you feel sluggish and thought-out. So drink your coffee or tea a few minutes before you give your talk and not just before the meeting starts. When you run low on NE you run out of mental gas. This shows how mind depends on meat in the short term. Learning is a long-term change in those same release rates. It governs how much transmitter gets dumped per second.

So learning changes synapses. That still does not tell us how to encode or decode a pattern in a neural net. How or where do you store a pattern? It can't be in one brain cell or computer-simulated brain net. Each net has thousands or millions of neurons in it. And each neuron does nothing but fill up with electrical spike juice and send out spikes of its own. How do you put a pattern in that?

The answer is in whole fields of neurons. The neurons speak a language of spikes. The neurons translate sense patterns and mental patterns and muscle-contraction patterns and emotion patterns into its own spike language. We think of language as made up of sentences made up of words made up of syllables or phonemes. In brain language the syllable is a *pattern of activation*, a whole field or slab of neurons that reverberate or resonate.

Think of the letter S. That is a pattern and so is the letter E. Say we want to teach a neural net to learn the rule "If S then E" or just to associate the two patterns. Say the net learns the S-E pair. Right now don't worry how it learns the S-E pair. Just say that it learns. Then if we feed the net an S or even a part of an S, it should give us back the full pair S and E. To do this in 1986 I came up with a net I called a BAM or *bidirectional associative memory*. In later years I extended the BAM in many directions but to learn the S-E pattern the simplest BAM works. This BAM uses two fields of neurons. One field codes for S and the other codes for E. A web of synapses connects both fields in one great pile of spaghetti (Figure 11.2).

We then feed the S-E pattern into the synapses. This step uses a simple math trick called correlation. It just multiplies small S pieces by small E pieces. There are many more complicated ways to do this but for now we stick with the simplest. So grant that we have taught the S-E pattern to the BAM synapses. We have stuffed its synapses.

Now think of each neuron as a light bulb. Real bulbs and real neurons are fuzzy and turn on to some degree. Real BAM neurons do that too. For now say they are bivalent. Each bulb is

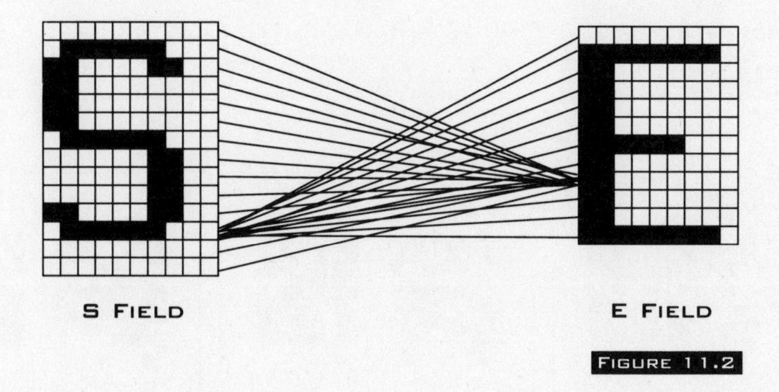

FIGURE 11.2

on or off. Each neuron fires or not, spikes or not. The net has two fields of neurons. Say the S field has 140 neurons—14 rows with 10 neurons in a row. Say the E field has 108 neurons—12 rows with 9 neurons in a row. So each field is a rectangle or grid of light bulbs. S and E might glow as a neural pattern (Figure 11.3).

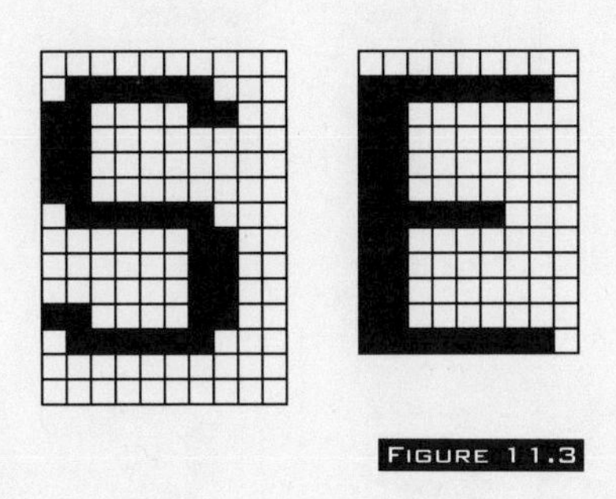

FIGURE 11.3

Now say we turn off some lights and turn on others. If we do this at "random," it will noise up the S-E pattern. Say this noise changes or "flips" 99 light bulbs or 40% of the net. Then the BAM net still tends to find the S-E pair in the noise. A noisy S passes over the wires and helps recall a noisy E but with less noise. The less noisy E flows back and helps recall a less noisy S, which flows back to clean up the E, which flows back to clean up the S, until the S and E are noise free. Then the BAM resonates on the S-E pair. The S recalls the E and the E recalls the S until you turn off the BAM or change its synapses (teach it something new). This is order out of chaos and associative recall.

FIGURE 11.4

Figure 11.4 shows eleven snapshots of the BAM recall process. This BAM has also learned the pairs G-N and M-V. The BAM stores all three pairs in the same memory matrix, one pair stacked on the other like diffraction patterns in a hologram. If you feed the BAM a 40%-noise G-N or M-V, it will recall the clean G-N or M-V pairs. If you put too many pairs in the BAM, it may fail to recall or clean up all the pairs. But it will resonate on *something*, some pair. The BAM will always "converge" or "stabilize." That is a theorem.

How does the BAM net work? How does each neuron know what to do to recall a learned pattern? No neuron knows. Each neuron acts like each house that gets calls and has no idea what the whole phone net does. Neurons are dumb and selfish. They just add up the on-off signals that flow into them and turn on when the sum is big enough and turn off or stay off when it is not. To use Adam Smith's market idea, each neuron acts as if an "invisible hand" guides it to recall the S-E pair though no neuron knows it should or that it helps do so.

Sound too good to be true? That's the power of neural nets. They *self-organize*. And maybe our brains act the same way. In 1977 Shun-ichi Amari of Tokyo University found the math that describes how nets of simple on-off neurons behave. He showed that the whole net converges or "cools" or "settles down" into a global equilibrium *as if a ball rolled down a well and stopped*. In 1982 John Hopfield of CalTech called these wells energy wells and showed that they minimize the "energy" of the net. He even showed how you could make a net find the lowest-cost solution of a scheduling problem or find one of the shorter tours a traveling salesman might take if he travels to each city once and then returns home. A BAM net does the same thing but it works with pattern pairs instead of just patterns.

The geometry looks like a sheet with bumps or wells on it. You can think of a pattern as a ball on the sheet. To store a pattern you dig a well for it and put your ball in it. Then a new pattern ball Q enters the neural net as if you dropped the ball on the sheet and it rolled into the closest well, in this case the P well (Figure 11.5).

You need no neural math to see why neural nets can classify patterns fast. The ball Q rolls into the P well and stops. Does it matter that there are other wells? No. There may be 100 wells or 100 billion. It does not matter. Q still rolls into the closest one and the time it takes to roll stays the same. That's why neural nets or *neurocomputers* differ from computers. In a computer the more you store the longer it takes to search for a match. Brains and neural nets don't work that way. The time it

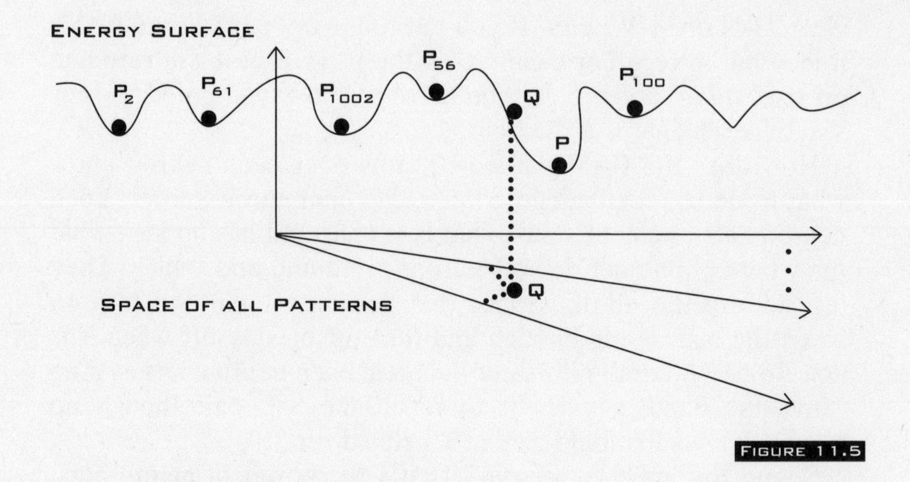

FIGURE 11.5

takes you to recognize your parents' faces in a photo does not change from age 5 to age 40 to age 70. But in that time you put more patterns in your brain.

On this view a *thought* is an energy well or point of resonance. *Thinking* is the ball falling into the well, the road to resonance. Stephen Grossberg of Boston University has shown that we learn a new idea or pattern only if it resonates with what we expect to see or hear or think. He has extended this "adaptive resonance" theory to the learning of fuzzy-set concepts.

Now the punch line: *Learning sculpts the energy sheet*. In the phone case above the phone lines stood for synapses that learn or change as calls flow through them. Real synapses work that way. We do not know the fine detail but we know you use them or lose them and the more you use them the bigger they get. The neural math shows that then you get an energy surface that changes as the synapses change. Learning is a shape change in the energy sheet. Each time you see your face in the mirror you learn it a little more, you dig a wider well for it on the energy sheet. The man in the iron mask goes so many years without seeing his face that he forgets what he looks like. That means his well has shrunk and gone flat. You practice to make perfect and to stay in shape. That keeps your wells well dug.

The energy sheet also gives a new view of *déjà vu*. We have all stood in a new market or new shop or sat with a new friend or read a new page and felt that we have done it before. It feels as if this new experience is just a repeat experience, an experience we have had before, have learned before. These *déjà vu* moments may be "spurious" wells on our energy sheet. Real experiences dig wells but they can also dig a few extra wells in the process. These are junk wells, leftovers from the learning process, like the junk heaps left at a construction site. When a fresh event comes close to one of these stored junk events, the ball rolls down the junk well and resonates at the bottom. Then we feel that we recognize or recall the fresh event.

We can also dig a well at the center or average of many experiences. The experiences are old and "real" but their average is new. In the spirit of Plato we can dig a red well when we see red blood and red apples and red toys. Good and God and the lovely may be just fill-ins or dig-ins or averages that we never experienced but only "abstracted" from clustered data. We generalize as our neural nets dig wells in our energy sheet. Like *déjà-vu* wells these make *new* ideas.

Here's one more speculation on the energy sheets in our nets in our head. Near-death experiences. These fall into two clumps. The first clump deals with seeing a bright white light. This may be the light that shines on the operating table or some large adrenal flash in the brain. The second clump deals with claims that just before death or in a near-death fix "your life passes before your eyes." That has about it a sense of poetry and irony. Show the poor fellow the richness of his life and then kill him. I think natural selection and neural nets can explain this last picture show.

Your mind knows when you are in a tough spot, when the Mack truck comes right at you or you fall from the rail and see the street racing up at you or you hear the guillotine blade drop. Fight-flight-fright adrenals take over. Your genes and the mind that sits with them and obeys them do not want you to die. The wire is tripped and the adrenals flash in one all-out associative search for something that will save you. *Massive associative search*. Red alert. Open the emergency doors. Search all the stored data nets and search them in parallel and search them *fast*. Check all the energy wells at once. The massive search would fill your mind's eye. You would not see your life play in front of you as a film plays in a theater. It would not be serial but parallel. Your near-death fix would trigger thousands of old stored events that somehow resembled it. The threat event would act as a ball or whole set of balls that roll down hundreds of energy sheets in hundreds of neural nets. Related past events and events related to them would all flash in your mind's eye at once. From this you might find the right thing to do to get out of the fix. It might just "pop" into your mind. Or the fix might pass on its own and after that you would just remember the massive associative flash and tell stories about it. If this view is right, you might recall it the next time you get the big-fix flash or right after you get the flash. It's just something to think about.

The big point is that learning changes an information medium. In neural nets the medium is an energy sheet. In us it is a web of synapses or webs of synapses or webs of webs of synapses and maybe some muscle-contraction rates. The information medium can be anything. A computer learns when you change its software or memory circuits. Warm wax learns your palm print when you push your hand into it. The canvas learns the painting when the artist smears paint on it. Even your lawn of green grass can learn what you mow it. This odd case gives a good example of learning in a parallel information medium.

The grass blades learn what they are cut. They grow in parallel as neurons fire in parallel and they act as a plastic information medium. You can throw out hundreds of grass blades and still read your name that you mowed in the lawn. You may need to stand on a roof to read it. You could encode or mow all known information in a big enough lawn. The lawn forgets what you teach it if you do not remow the lawn to the same shape. In the same way a neural net loses its energy wells if you do not retrain the net with the same or similar patterns. Without practice it all goes to seed.

LEARNING RULES

1: RULE = ENERGY WELL

Neural nets can dig energy wells for fuzzy rules. A rule is a pattern. A rule says *IF this THEN that* just as a BAM pair says *IF S THEN E*. As in the FAT theorem, a rule is a patch. So tie energy wells to patches. Let experts dig our wells with their examples. Key idea: *Fuzzy Rule = Neural Equilibrium*. Same idea: Rule = Energy Well.

I saw this many years ago. During the day I worked in neural nets and wrote papers on them and taught courses and seminars on them. At night I kept up the fuzzy research. The government paid me a lot of research money to work out the neural math but paid me nothing for fuzzy math. I always had to sneak it in on the side. In 1987 I wrote my Ph.D. dissertation on pure fuzzy math and that same year wrote several neural papers and chaired the first international neural conference and got my job at USC to be a "neural expert." At USC the department folks used to take the term "fuzzy systems" off my page in the faculty brochure. They put it back a few years later when the fuzzy money started to flow but that is another story. I worked in two fields that I loved and early on I started to mix the two. In 1983 I came up with the fuzzy cognitive map that we will look at in a moment. In the 1980s I worked out the math of how to grow fuzzy rules from neural nets and data. Then my graduate students and I applied the theory to smart tasks like backing up a truck and trailer or shooting down a cruise missile or compressing image data or TV scenes before you send them.

The idea is that rules make clusters in the data and clusters are candidate rules. Data clusters, like galaxies of stars and the cell clusters that make up nerve or muscle tissue, are structure. Clusters are patterns. They hold information while they last.

So rig or "train" with samples a neural net to find the clusters as it equilibrates. That means data clusters dig their own energy wells. The more examples the marine gives you of how to shoot down a SCUD missile, the wider and deeper the rule wells he digs on the energy sheet. The more examples the better. Data in, rules out. The next section shows a second and related way to use neural nets to grow rules from clusters.

LEARNING RULES
2: Rule = Data Cluster

Fuzzy rules are patches. We saw that with a fuzzy air conditioner. The next step ties patches to data clusters. This gives new force to the FAT theorem that says a fuzzy system can model any system by covering its system curve with small fuzzy patches. With enough data the fuzzy system can *learn* any system.

Adaptive fuzzy systems work at two levels. Each level approximates something. At the small or local level a neural net approximates patches or rules. At the big or global level the patches approximate the whole system. That is a nice one-two punch.

Look at the fuzzy air conditioner again. It used five rules and so the system has five patches (Figure 11.6). Each point in the temperature-speed square is a pair of numbers like 65°, 50. The first number is a temperature. The second number is the speed the air motor runs at that temperature. Each patch covers many points. In fact each patch covers infinitely many points if we allow arbitrarily fine degrees of temperature and motor speeds.

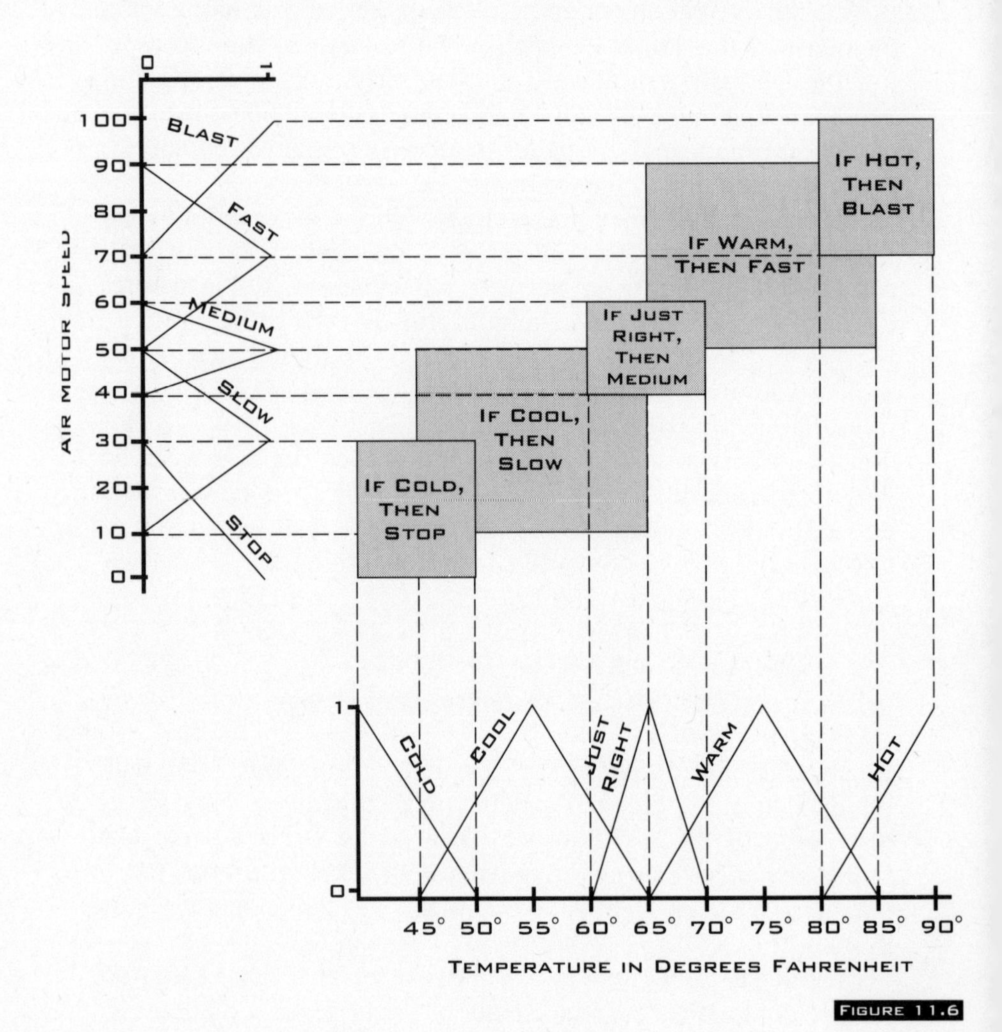

FIGURE 11.6

In practice there are only so many temperature values and motor speeds. There are only so many possible pairs like 65°, 50.

Points make up patches just as feathers make up a pillow. A patch will grow if we add more points in a region, if we make it denser, if we form a cluster of points. More points help the neural net dig a wider energy well over the patch.

Say we watch a human expert run the air conditioner. We do not know what rules she uses and she does not know. She just adjusts the motor speed when the temperature changes and she does it well enough so that we call her an expert. Every minute we write down the room temperature and where she has set the motor speed. Say at first it is 65° and she sets the speed to 50. So we write down the pair (65°, 50) and plot it as a point in the temperature-speed square. Someone opens a door and hot air fills the room. At minute two the temperature is 67° and she sets the speed to 53. We plot the point (67°, 53) in the square. At minute 3 we read 70° and she sets the speed to 60. We plot the point (70°, 60). We keep this up for hours. In time the room temperature may fall. Clouds roll in outside or night falls. Then we might read 61° and she might set the speed to 41. In a few hours we have filled the square with a swarm of points and stuffed one pillow (Figure 11.7).

The swarm is a data cluster. It shows how an expert answers questions or associates outputs with inputs. As David Hume saw, it shows how we associate similar outputs with similar inputs. It defines a rule. In a neural net it digs a deep wide energy well as a sidewinder digs a pit to hide in in the desert sand.

More data pairs lead to more clusters and so more energy wells and so more rules. I call this learning scheme *product space clustering*. It just means that a rule is a cluster in the data. More clusters, more rules. If we vary the temperature from 45° to 90°, our expert must show how expert she is over many more cases. This tends to lead to several clumps in her data and these are more rules (Figure 11.8). Real data clusters need not fall so neatly into the patch cells defined by the fuzzy-set triangles. In practice engineers have to move and reshape the fuzzy sets to tune the system and to better fence off rule clusters. Or we can use "error" balls or ellipsoids around some data points to tell us where the patch borders lie and thus how to tune the fuzzy set along the edges.

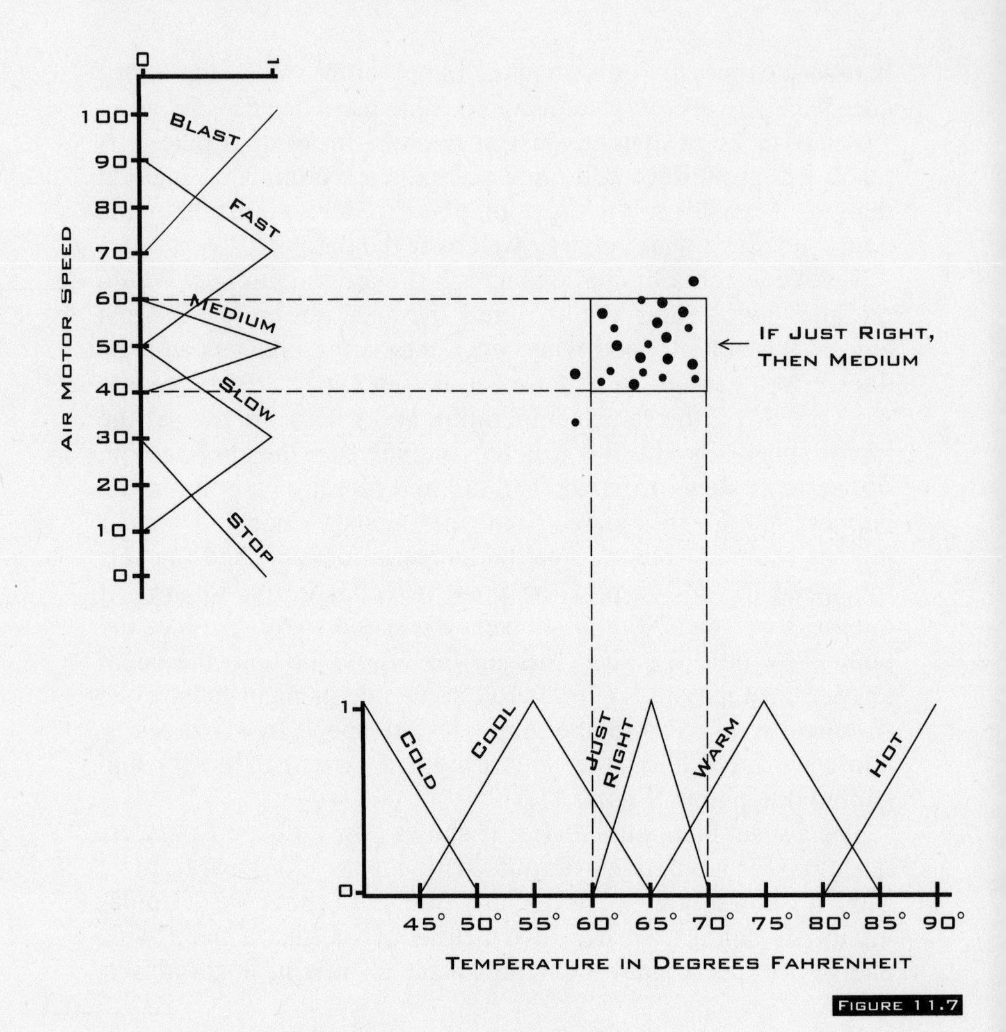

FIGURE 11.7

Neural nets can learn rules without energy wells. In some sense the energy wells are always there but they do not dictate how we design the neural net. These nets are *adaptive vector quantizers* or AVQ nets.

Each neuron has its own web of synapses that flow into it. The web defines a point, or AVQ dot, in the system geometry or "space." The neurons "compete" as each new piece of data rolls in. The new data also define a point in the geometry. A neuron "wins" if its AVQ point is closest to the data point. Then it gets to learn by changing its synaptic web a little so that it looks more like the data. This means the winning AVQ dot

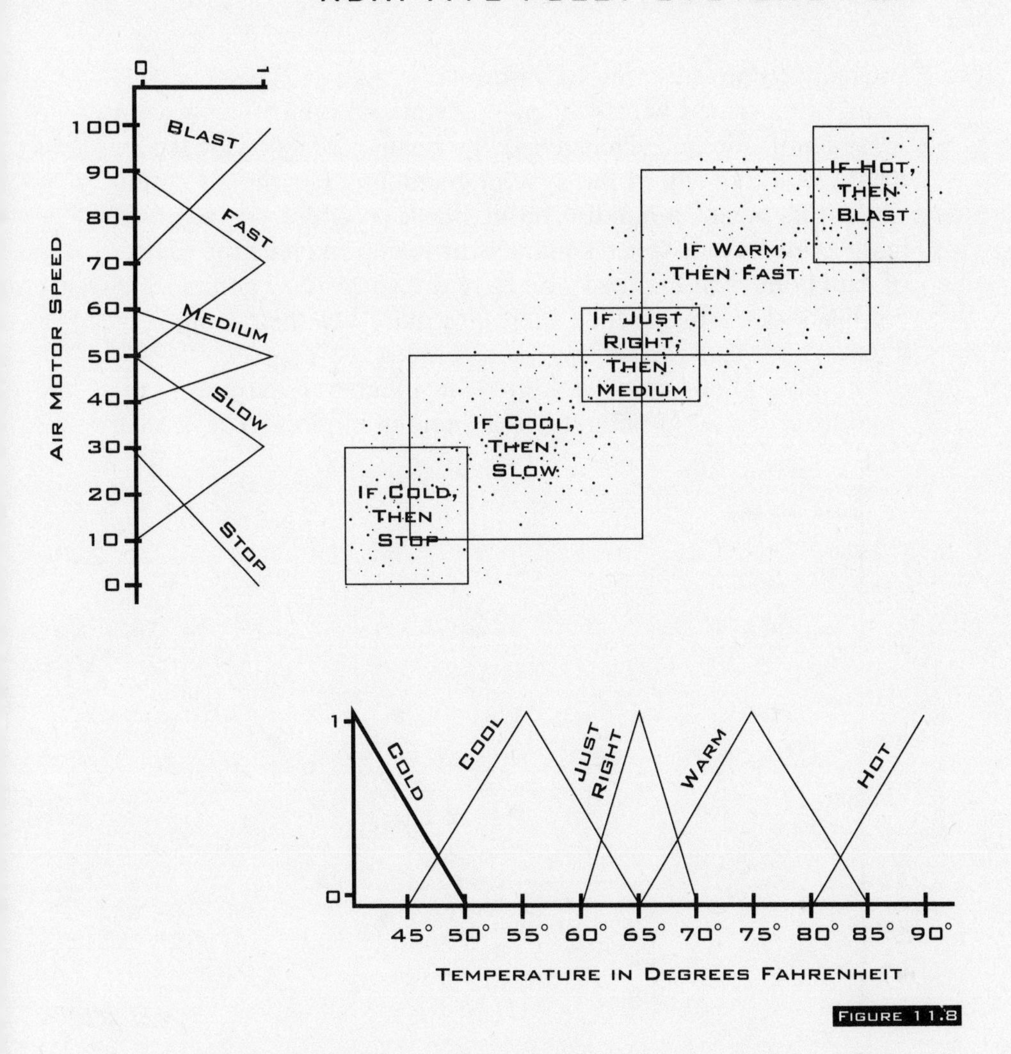

FIGURE 11.8

moves a little closer to the data point. A new data point comes in and the process repeats: Pick a winner. Change a web. Repeat.

You quantize when you round off or when you pick an example. Imagine a speech system for your car. It must recognize words like "open," "close," "slow down," "too hot," "too loud." The speech chip stores a template or prototype pronunciation of each word. There are over a million ways to say "open." The chip stores just one and matches your words against it. The word it stores quantizes or stands for the whole speech class of "open." So it had better be a good one.

Quantization compresses information. An AVQ net slowly changes its stored word "open" as you and others speak your versions of "open." Change slowly means learn slowly. It also means *move* slowly in the system geometry. For the air conditioner this means a handful of big black AVQ dots move about in the temperature-speed square so as to approximate the scatter of data points. In Figure 11.9 I have drawn 25 AVQ points as 25 big black dots. In practice there are more but their number is fixed.

The data points grow and grow in number. Learning never stops. But the AVQ dots stay the same in number. They move

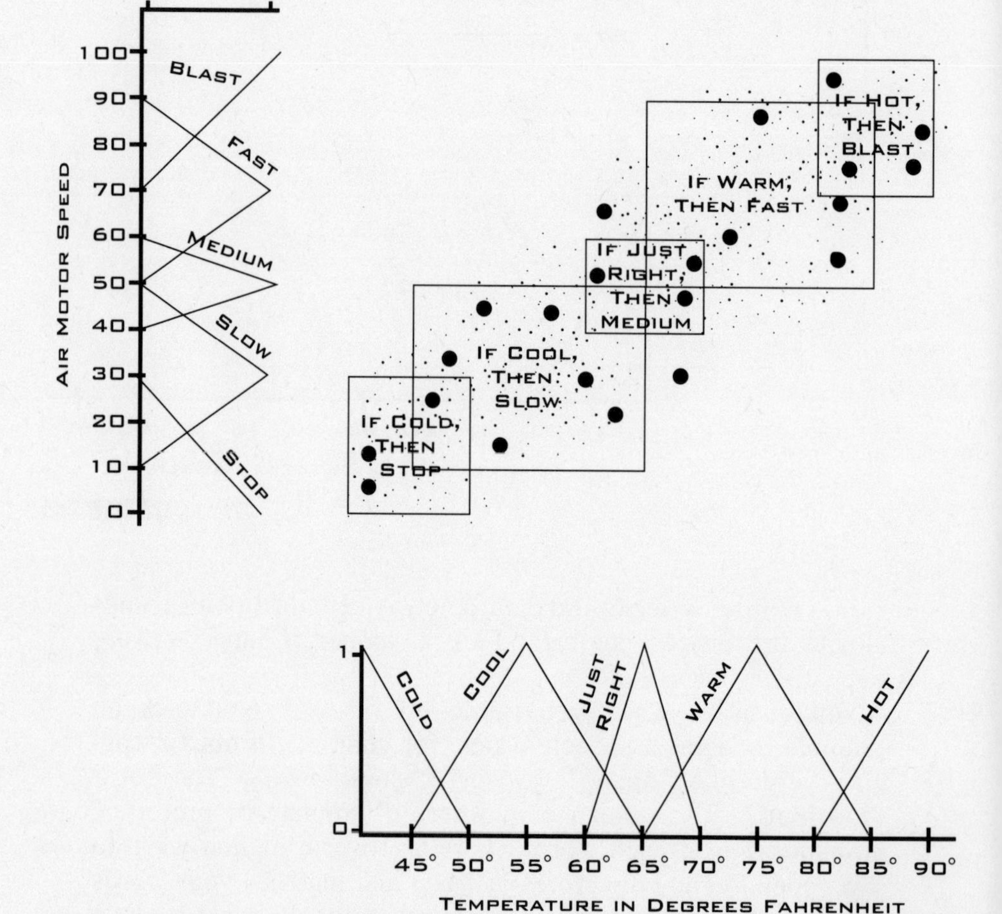

Figure 11.9

around as new data points pour in. Figure 11.9 gives a snapshot of AVQ action. The big AVQ dots tend to be dense where the data points are dense and sparse where the data points are sparse. This means the AVQ dots estimate data clusters. So they estimate fuzzy rules, since rules are patches and clusters cover patches.

So how do you learn rules with an AVQ net? You just count the big black dots in the rule cells. If a cell has an AVQ dot in it, add that rule to the system. You can weight the rules this way too. Some cells have more AVQ dots than other cells have. This can mean they are more important to the expert or it can just mean that the expert hangs out in that region of the plane. In practice we often require two AVQ points per cell to count as a rule. This cuts down on bad rules that can come from noisy data or from a rare bad example. Even the best experts make mistakes.

Each AVQ dot lies at the center of its own "error" or "covariance" ellipsoid. The size of the ellipsoid depends on how large a region the AVQ dot stands for or quantizes. Regions of sparse data give rise to large ellipsoids and thus large patches and thus less-certain rules. Tightly clustered data give rise to smaller ellipsoids and thus more-certain rules. We can project the overlapping ellipsoids onto the edges to both find and tune the first set of fuzzy sets. This method picks both the size and number of the first set of fuzzy rules.

An AVQ net fills in the DIRO black box: Data in, rules out. It slowly sucks the expert's brain as she feeds it expert data. At first the AVQ points do not spread out well. That gives a small number of bad rules. Then with more training data the AVQ points spread out to track the data. That gives more and better rules. More data refine the spread of AVQ points and polish the rules. At this point we can drop the expert and just work with the fuzzy rules. They will give the same behavior. In practice we then polish these rules by playing with the size and position of the fuzzy sets or by letting a "supervised" neural net tune them. Then the tuned fuzzy system can outdo the expert. The student beats the teacher.

There are many ways to grow fuzzy systems from data. They all boil down to clustering data into rules. So I have focused on that idea and ignored a few dozen other approaches that in the end do the same thing. These make up the new field of adaptive fuzzy systems, or neuro-fuzzy systems, or fuzzy neural systems. The neural nets just use data to find or tune rules. The rules are patches and they cover a wiggling system curve in accord with the FAT theorem.

But there is a fuzzy net that differs from FAT systems. It is as much a web of fuzzy rules as it is a web of neural synapses. You can use these fuzzy nets to model events in politics and history and medicine and military planning. They are fuzzy cognitive maps or FCMs. You can't forget the acronym: FCM rhymes with "buck 'em."

FUZZY COGNITIVE MAPS: FUZZY PICTURES OF THE WORLD

A *fuzzy cognitive map* or FCM draws a causal picture. It ties facts and things and processes to values and policies and objectives. And it lets you predict how complex events interact and play out.

An FCM stands behind every op-ed article and every political speech. Here is one of the first and simplest FCMs I ever drew. I had read an article by Henry Kissinger* in the Sunday paper and then turned it into a Mideast FCM. Each arrow in Figure 11.10 defines a fuzzy rule. It defines a causal link or connection. A plus (+) means causal increase. A minus (−) means causal decrease. The plus rule in Figure 11.11 means that if Islamic fundamentalism goes up then to some degree Arab radicalism goes up too, and if Islamic fundamentalism goes down then to some degree Arab radicalism goes down too. The minus rule means that if PLO terrorism goes up then to some degree the strength of the Lebanese government falls, and if PLO terrorism goes down then to some degree this strengthens the government (Figure 11.12). You can weight these rules or arrows with any number between 0 and 1 (or between -1 and 1) or you can use word weights like "a little" or "somewhat" or "more or less." That's part of what makes an FCM fuzzy.

---

*From Henry A. Kissinger's essay "Starting Out in the Direction of Middle East Peace," *Los Angeles Times*, 1982. This FCM appears in my article "Fuzzy Cognitive Maps," *International Journal of Man-Machine Studies*, vol. 24, 65–75, January 1986.

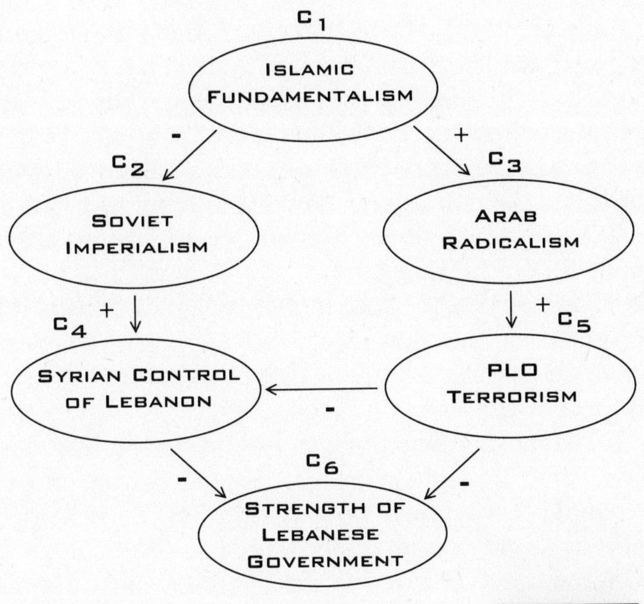

FIGURE 11.10

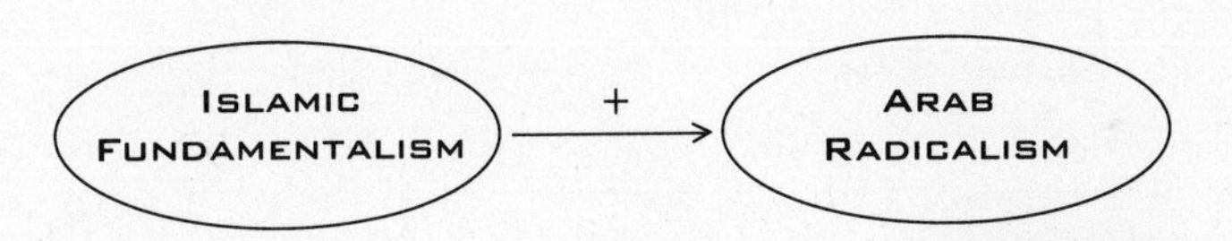

FIGURE 11.11

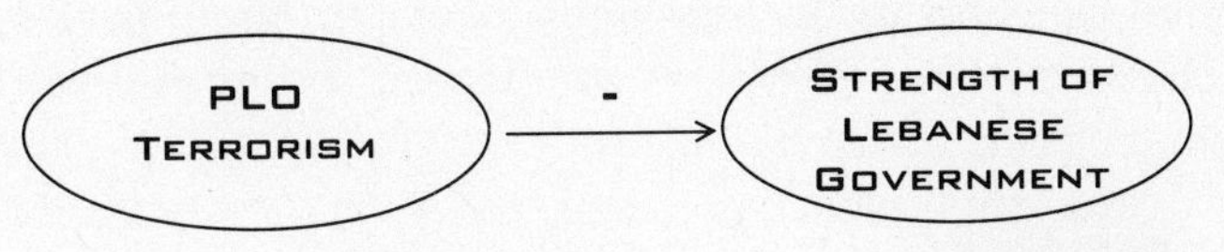

FIGURE 11.12

The nodes or concepts are fuzzy too. Each node can fire to some degree from 0% to 100%. Each node is a fuzzy set. Every event fires it or belongs to it to some degree, usually to zero degree. In the simplest case the nodes are on or off. They fire or turn on if we see an act of PLO terrorism or Islamic fundamentalism or if enough causal juice flows into them to trip a threshold. In real FCMs the nodes are fuzzy and fire more as more causal juice flows into them.

In the Kissinger FCM ISLAMIC FUNDAMENTALISM is a policy node. No causal arrows flow into it. You or God or a government controls it, turns it on or off, or fires it to some degree. It lets you play what-if games.

What if Islamic fundamentalism goes up in the Mideast? The FCM says Arab radicalism will go up and Soviet imperialism will go (would have gone) down. This lets loose two causal chains. First, as Soviet imperialism falls so too to some degree will fall Syrian control of Lebanon—perhaps since the Soviets had helped train and equip the Syrian army. But a fall in Syrian control strengthens the Lebanese government. Second, more Arab radicalism leads to more PLO terrorism and this has two effects. It directly weakens the Lebanese government. And it decreases Syrian control of Lebanon, which strengthens the Lebanese government. So these two effects may cancel. Or one path may outweigh the other. It depends on the fuzzy weights. All three paths flow from the top node. Two of the three paths strengthen the Lebanese government and one path weakens it. So with equal weights the FCM predicts that an increase in Islamic fundamentalism will to some degree strengthen the Lebanese government.

You can argue with how I drew the FCM map from Kissinger's article. Or you can argue with the article. You may want to add other concepts or delete some of these. Or you may draw other arrows or rules or swap their signs or weight them in some new way. That's fine. An FCM lets you do that. It lets everyone pack her own wisdom and nonsense into a math picture of some piece of the world. But once packed in, the FCM predicts outcomes and we can compare these with data to test them. The outcome of large FCMs may surprise you. The best most of us can do is argue about single arrows. We do less well when we try to reason with a large set of connected concepts. FCMs help us see the big picture and do something with it. Best of all behind the FCM lies a pure math scheme that computers can use and we need not worry about.*

I came up with FCMs in 1983 to solve a problem. Cognitive maps of one type or another had been around for ten years in psychology and political science. I was working on my Ph.D. and worked in aerospace to pay the bills. That was as close as I ever came to military experience and for me it was too close.

I needed a "smart" way to model the target value of a bridge in the middle of a battle. The old cognitive maps did not let you have fuzzy nodes and fuzzy arrows or rules. And they did not let you have *feedback*. The arrows could not form a closed loop as they do in the real world. If you poured causal juice in at the top of an FCM—if you turned on one or more nodes—then the juice flowed straight down and ran out at the bottom. It could not swirl round and round. The FCM was not a *dynamical* system. After all how could you make sense of FCM with "cycles" or closed loops?

FCMs thrive on feedback. That was the key. I was trying to work with FCMs as experts in artificial intelligence work with computer programs, with long chains of if-then rules. The new FCM idea was to drop that and view the FCM as a neural net. Make it a dynamical system. Let it swirl and reverberate like a Bidirectional Associative Memory. It will settle down. It will converge. It will equilibrate. Figure 11.13 is one of the first feedback FCMs I drew.

This FCM shows how bad weather can affect how fast you drive on a California freeway in the daytime. The arrows or fuzzy rules have word weights like "usually" and "a little." The FCM has two small tangles of feedback. It has two short cycles. In the first feedback loop freeway congestion somewhat increases auto accidents. And auto accidents often increase freeway congestion. That's feedback: back-and-forth flow of information and influence, two-way causality. In the second closed

*The math scheme writes the state or snapshot of the FCM as a list of numbers, as a vector. The bit vector (0 1 1 0) means that at the moment of measurement an FCM with four nodes has only the second and third nodes on. Fractions or fit values measure the degree that nodes are on. The math scheme writes the causal connections between the nodes as a large square matrix of numbers. If there are n nodes in the FCM, the FCM matrix has n^2 numbers in it.

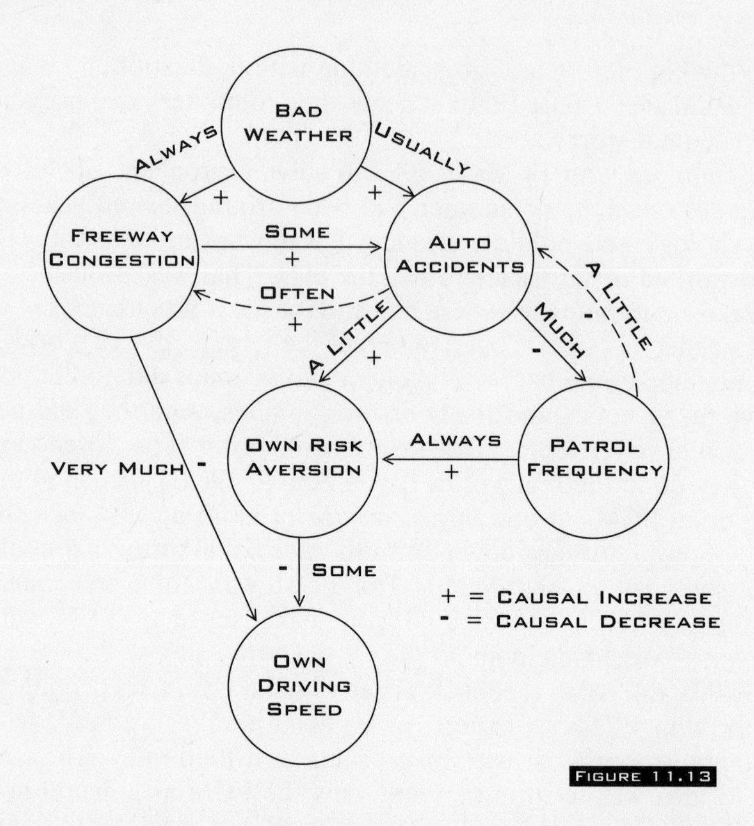

FIGURE 11.13

loop auto accidents much decrease the frequency of highway patrolmen on the road. But more highway-patrol frequency helps decrease auto accidents.

Feedback loops grow as you add more nodes to the FCM. Consider the FCM of South African apartheid politics in Figure 11.14.

What can you do with all those closed loops? You can't walk through this FCM as if it were a tree (has no closed loops) as is the Kissinger FCM. You would go round and round in infinite loops. You can argue with each link. Maybe you don't like an arrow's direction or you want to change the arrow from minus to plus or from plus to minus or you want to take the whole arrow out. But even if you agree with the FCM links, what can you do with the thing? Feedback and accuracy seem to oppose. Accurate pictures of the world are full of feedback. But how do you work with feedback?

You look for *hidden patterns* in the FCM edges. You turn the

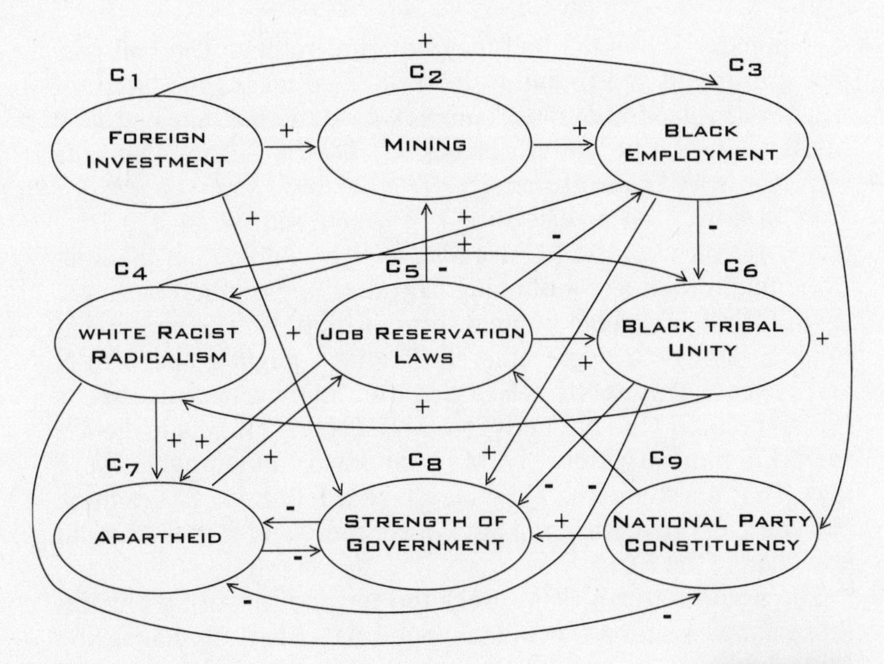

FIGURE 11.14

FCM net on and let it swirl. You let the dynamical system converge to an equilibrium. Events blink on and off or fire to some degree. One concept or event swirls into others. Mining up makes black employment go up. That makes black tribal unity go down but makes white racist radicalism go up and helps strengthen the government. More white racism makes for more apartheid feeling or laws and it increases black tribal unity. This in turn increases white racism. And round and round it all swirls as the hidden pattern emerges.

What is a hidden pattern? It is where the swirl stops. It is an equilibrium. Look at the BAM energy sheet again (Figure 11.5). A hidden pattern is what ends up at the bottom of the energy well. In the BAM case the system state rolls as if it were a ball on the sheet and rolls down the closest well and stops at the bottom. The well is an *attractor basin*. That is what dynamical systems theorists call it. The point of rest at the bottom is the *attractor*. It "attracts" the system ball to it. BAMs always stop at the bottom of the well. That means the attractor is a *fixed point*. The BAM resonates on the fixed S-E pair. In other dynamical systems the ball may not stop rolling. The ball may just spin round and round in the well. That means the attractor is a *limit cycle*. In the most complex case the ball moves round in the well with no pattern or period. That is "chaos" and the attractor is a *chaotic* attractor.

FCM balls stop or else they roll round and round in a limit cycle. If you view the FCM nodes as light bulbs, then the limit cycle might look like a blinking light message on the Las Vegas strip. One light snapshot might form or blink N. The next might blink E. The next might blink T. The next might go back to N and repeat. This spells NET. But the limit cycle really spells NETNETNETNETNETNETNETNETN. . . . It gets locked in the NET hidden pattern. FCMs cool down to limit cycles very fast. That's a theorem. They may take a few steps to get there but you can prove they will always get there and stay there until you change the FCM.

The South African FCM looks too tangled to work with. But those tangles make the hidden patterns richer. It makes the FCM dynamical system more dynamic. It tends to give the FCM more attractors. I drew this FCM based on an article by economist Walter Williams* who had just come back from South Africa in 1986. Then most newspapers were calling for the U.S. to "disinvest" in South Africa to pressure the Botha government to abolish apartheid laws. The idea sounded good. But Williams argued that it would hurt the very persons it meant to help. The poor would have fewer jobs and a lower standard of living and that would destabilize the country. I thought that was a clever claim and wanted to see if an FCM would say the same thing. If it did, it would not mean that Williams was right. It would just mean that all his little arguments and assumptions added up to support his conclusion. He painted his causal picture a sentence at a time—a node or two and an edge at a time. All those nodes and edges make up and hide the global patterns in the system.

I drew the FCM and then ran it on a computer. Each node could turn on or off. To test the claim I turned on the Foreign Investment node and kept it on until the FCM cooled down and fell into an attractor. The FCM stopped in a fixed point. That

---

*Walter E. Williams, "South Africa Is Changing," San Diego *Union*, Heritage Foundation Syndicate, August 1986.

was the hidden pattern or the FCM prediction or the answer to the question What if countries invest in South Africa to a high degree? In this pattern some nodes were on and the rest were off. On were FOREIGN INVESTMENT, MINING, BLACK EMPLOYMENT, WHITE RACIST RADICALISM, STRENGTH OF GOVERNMENT, and NATIONAL PARTY CONSTITUENCY. APARTHEID and JOB RESERVATION LAWS were off. The answer is qualitative. It is a pattern prediction not a number prediction. So far it seemed to support Williams's claim. At least sustained investment did not lead to social chaos.

What if foreign countries "disinvest" in South Africa? How do you test that on an FCM? You do not start out by turning off the FOREIGN INVESTMENT node. You start out by turning it on as I had done and you keep the node on. Once the FCM converges to its hidden pattern then you turn the node off. FOREIGN INVESTMENT is a policy node so no causal juice flows into it. Once off it stays off. So I turned off FOREIGN INVESTMENT. The FCM swirled through many states and then stopped on a two-state-limit cycle. It blinked back and forth between two states. In one state only WHITE RACIST RADICALISM and JOB RESERVATION LAWS were on. The government had fallen. In the other state only BLACK TRIBAL UNITY and APARTHEID were on. This hidden pattern has the flavor of social chaos and a race war.

The test supported Williams's claim. It did not prove the claim. FCMs can't do that. They can help show how the whole of your beliefs behaves. They show *global* patterns. As the FCM grows more complex the hidden patterns become harder to see. You need a math method to find them. And they may surprise or offend you. This FCM so offended one reviewer of my paper that he demanded that the editor take out the names APARTHEID and FOREIGN INVESTMENT and the names of the rest of the nodes. The reviewer did not like Williams's politics or world view and neither did the editor. So when the article ran there was the tangled FCM with node numbers instead of concept names. The analysis talked about node 6 and node 7 on the other nodes off. It made no sense. Academic politics and the pursuit of truth can come out that way.

I like the fact that FCMs can offend and can paint causal pictures of hot topics. They lay bare your beliefs and biases and

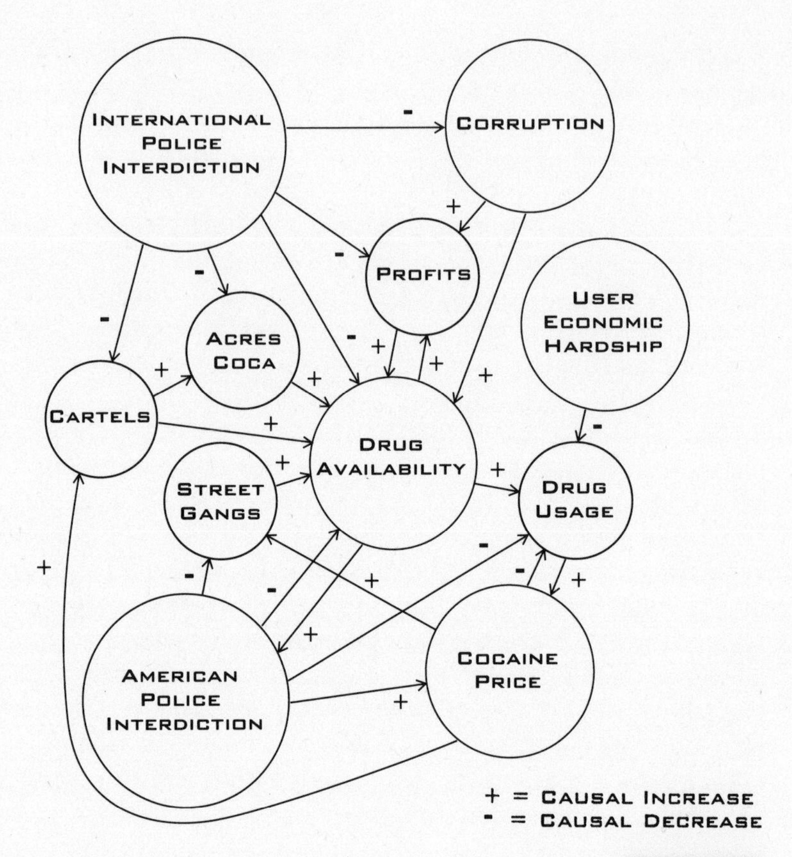

FIGURE 11.15

your grasp of the world. They help most in the value clashes that mix head and heart. I have found that students draw their best FCMs when they try to model a clash of governments or the spread of AIDS or any topic that deals with power or sex.

Dr. Rod Taber at the University of Alabama at Huntsville has published FCMs on medical and social problems like this FCM on the cocaine market in Figure 11.15*.

On a hike in the Mojave Desert I asked big burly Dr. Rod about this FCM. I tell you this because it shows how to tune an FCM and how hard that is and how good it would be if a neural net could tune it for you. It was in winter break and the wind blew snow pellets at us at almost right angles. We were coming

*Rod Taber, "Knowledge Processing with Fuzzy Cognitive Maps," *Expert Systems with Applications*, vol. 2, no. 1, 83–87, 1991.

down the Providence Mountains to beat the storm before night fell. The cocaine FCM was something to talk about to forget about the pack load and the cold wind that cut through us in blasts. I looked at the ground to keep the snow out of my eyes. The big white snow pellets stuck to the orange rocks and gravel and did not melt and looked like small-curd cottage cheese spread out on a pan of yams.

I asked Dr. Rod if his FCM said that the cocaine supply would fall if the U.S. stepped up its War on Drugs. I said it had better. If not he would have to add the right fuzzy arrows to fix it. He did not want to change the FCM since the paper was in galley proofs and ready to appear but he would change it if we could find the missing links.

The first task was to see if the cocaine supply went down or shut off if you turned on the node AMERICAN POLICE INTERDICTION and held it on. We had a copy of the paper with us but that was all. To test the claim we had to work with matrix algebra and an FCM matrix that had 121 (11×11) numbers in it. We had to multiply and add numbers in our head and then see if a node turned on or off. Each step was like multiplying 235 by 9478 in your head while we walked and slid down a desert mountain in a snowstorm. Mozart would have had no problem with it but we struggled. By the time we got into our sleeping bags we had found that the FCM did not say the right thing. Cocaine supply did not fall when the U.S. cops rose. We tested this later on a computer and it agreed with us. When you think of sleeping bags in the Mojave Desert you think of snakes crawling in with you. But there were no snakes in this cold. The rattlesnakes all slept interwined by the hundreds deep down in cave cracks and old jack-rabbit holes.

The next task was to find an edge that made sense and that made the cocaine supply go down when the cops went up. Dark fell and the snow still blew and we were not tired yet. I unwrapped a green light stick and bent it into a horseshoe. It glowed a bright phosphorescent green. I wedged it behind our heads in a creosote shrub. We had to hold the FCM paper close to the green light to see it and hold it tight to keep the wind from blowing it away.

What would halt or reduce the flow of cocaine to the U.S.? I argued that U.S. police alone could not halt the supply. They might dent it for a while but in the end they could only make the black market blacker and make the price higher and the profit higher—high enough to make up for the dent. And worse than that. Economics says that for a *fixed* demand as the supply falls to zero the price and hence profit rise to infinity. In this extreme the cops and soldiers lay down their guns and plant coca shrubs or make coca paste or ship or sell cocaine. But that was an extreme what-if case. In the short term the U.S. can keep the supply down if they use enough cops *and* if other countries help them. In terms of Taber's FCM you need both U.S. *and* international interdiction. That would keep down the supply.

So what was missing? The FCM needed feedback from the U.S. to get the local countries to help. Prices can send strong feedback signals. The FCM needed a plus path from COCAINE PRICE to the South American cocaine CARTELS. Price up means profits up. That could only help the cartels. Cartels up means local trouble up and U.S. calls for help up. To add this new plus link meant replacing a 0 with a +1 in the big FCM matrix of 121 numbers. Then we had to fight the driving snow and dark to redo the math. This time the FCM did what it should. Cocaine supply fell when the U.S. police effort grew. Later a computer confirmed this.

One link changed the output. One causal arrow changed the hidden pattern. Dr. Rod and I did not see this at first. We looked at the FCM and argued about it and tried out lots of ideas and tricks. You might think we could use a computer to test out all possibilities. You might try that with a chess game too. There are too many possibilities. There are vastly more possible chess games (over 10^{120}) than particles in the known universe (around 10^{80}). We rely on luck and feel to come up with new ideas. A computer relies on brute force as it searches branch after branch on a huge decision tree. Both ways are hard and hit their limits fast.

That cold night in the Mojave the blizzard moved in with gusts up to 70 m.p.h. It was too early and too cold to sleep in our icy sleeping bags and the FCM talk had long since run out. I would drop off to sleep and wake again in a few minutes when some toes or fingers froze up or a gust blew snow into my bag. This went on for twelve hours. At dawn the winds fell and we got to sleep at last. We got up in three hours and broke camp and rolled up our sleeping bags and took off our gloves to tie up the bags real fast in the zero cold. Then we saw the black cave hole just a few yards from us. We walked up to the cave and looked in. The cave went back for several yards and it was dry and stayed about 50° F. inside. It was a new kind of desert oasis. The night before in the dusk and snow we had missed the cave. And of course we did not think to look for a cave to sleep in. We had missed the shortcut.

Neural nets give a shortcut to tuning an FCM. The trick is to let the fuzzy causal edges change as if they were synapses in a neural net. They cannot change with the same math laws because FCM edges stand for causal effect not signal flow. We bombard the FCM nodes with real data. The data state which nodes are on or off and to which degree at each moment in time. Then the edges grow among the nodes. When I first tried this in 1985 I let the edges grow as synapses grow. Synapses correlate neural firings. But in an FCM that leads to spurious causal links. Two concepts do not affect each other just because they are both on at the same time.

So I once again turned to my "worthless" philosophy degree to look for ideas. Philosophers have said a lot about cause and effect. Until David Hume most believed in it. After David Hume in about 1800 they no longer believed in it. Hume said cause was an illusion, a "sentiment" the mind imposed on the flow of events, a pattern of correlation, a "constant conjunction of events." You could replace "Fire causes heat" with "If fire, then heat." I did not like this use of correlation. It was too static. Causality deals with changes.

I turned to John Stuart Mill, the heir to Hume, and his view of causality. In his *System of Logic* Mill said causality is "concomitant variation." We infer or feel or make up a causal connection between A and B if they *vary* or change together. A and B have to move in the same way. In neural nets the Hume or correlation view means that you multiply A by B to find the synaptic value between them. They call that Hebbian learning after neuroscientist Donald Hebb, who first wrote about it in 1949. I wanted to multiply how A changes by how B changes. In calculus the derivative measures change in a thing or process. Velocity is a derivative or change in position. Acceleration is the derivative of velocity or the change in the change of position. You don't need to know what a derivative is to see my point. You just need to know that a derivative, like a speedometer, measures change. So in 1985 I multiplied the derivative of A by the derivative of B and called it *differential Hebbian learning*. I put forth DHL as a way for FCMs to learn from data. But I also thought it might apply to real synapses in brains and some researchers have since suggested that it does. Right now I want to stick with FCMs to show how DHL works.

Let A and B be FCM nodes. A or B can be a social policy or the strength of a government or the target value of a bridge. Suppose A goes up a little. Then the difference between the new A and the old A is positive. So the change of A is positive (+). Now say B goes up a little too. So the change of B is positive (+) too. Then plus times plus equals plus ($+1 \times +1 = +1$). The friend of my friend is my friend. So you add a little number to the edge from A to B. Now say A falls. Then the change is negative (−). Say B falls too. So the B change is negative (−) too. Then if you multiply the two changes you get a positive value ($-1 \times -1 = +1$). The enemy of my enemy is my friend. So again you add a small number to the edge. So if A and B change together, we guess a causal link between them.

The reverse happens if they move in opposite directions. Say A goes up and then B goes down. Then multiply a positive by a negative to get a negative ($+1 \times -1 = -1$). The friend of my enemy is my enemy. The same happens if A goes down and B goes up. The enemy of my friend is my enemy. In both cases you subtract a small number from the edge and it tends to move to a negative link. If neither A nor B changes, you get zero and add zero to the edge. This gives you a small hint of what a computer can do when it runs DHL or some other learning law to create and tune large FCMs as data pour in. So now you know what you can do with a philosophy degree. Cash it into math.

Every time you change an FCM you learn in some way since learning is change. I worked out a scheme that lets you add one FCM to another (and you can weight the relative importance of each FCM). You can add up all FCMs. Plus edges can cancel negative edges. Or a lot of plus edges add up to a strong positive link and likewise for a lot of negative edges. It turns out you can write down an FCM for every book or article ever written and every speech ever given and add them all up into one giant FCM. That just starts the process. Then there are now many ways you can let the FCM learn for itself as a true adaptive fuzzy system.

Imagine this giant FCM cloud with all known human knowledge, with the good and the bad and the stupid. It might lie among a thousand linked computers. Or it might fit as a tiny chip in the back of tomorrow's laptop or in the base of your brain. News events fire its nodes. The latest learning laws turn the node firings into new causal links. The whole thing swirls and cools down to the latest prediction. Some FCM nodes fire entire sub-FCMs. Fuzzy-rule systems fire some nodes or sub-FCMs. Some nodes fire fuzzy-rule systems. Sub-FCMs break off and contradict the big FCM. Rival social factions might put forth their sub-FCM and shout its hits and explain away its misses. It's a fun idea. The point is the FCM framework lets you imagine it. It gives a new way to represent knowledge.

I think in the future we will see FCMs put at the end of opinion pieces or op-ed columns in the newspaper. Cut out the ego and cut to the FCM. At first the FCM will just be an appendix tacked on to impress us. Then we will get used to reading them. We will get better at checking the words in the article against the causal links in the FCM. The words will only argue for FCM nodes and edges. Then in time we will get fluent with FCMs. Journalists and lawyers and social scientists and mom and pop will feel safe with them. They will come to argue with them as in this century they came to argue with the charts and statistics of economics and medicine. Then the FCM will be the article and the words will go in the appendix. We may watch FCMs grow beneath the talking heads on news shows.

The fuzzy future will have many new tools in it. Fuzzy logic and neural nets will seep into the pop culture and into the rivers and streams of the Information Age. They will pass from threat to fad to tool. Fuzzy logic will change our world views in small ways and in deep ways. It will bring us closer to machines and bring them closer to us. And fuzzy logic will poke holes in moral absolutes. It will help solve some problems and will muddy up others.

Einstein's theory of relatively gave new force to the claim that "Everything is relative." Fuzzy logic and the machines that use it will remind us that "Gray's okay" and "Everything is a matter of degree." We may miss our days of the black and the white. But miss them or not they are gone.

IV

The Fuzzy Future

12

THE FUZZY FUTURE

Proposition 1. The world is everything that is the case.

Proposition 6.362. What can be described can happen too, and what is excluded by the law of causality cannot be described.

Proposition 7. Whereof one cannot speak, thereof one must be silent.

LUDWIG WITTGENSTEIN
TRACTATUS LOGICO-PHILOSOPHICUS

The fuzzy future will be full of smart gadgets. They will have high machine IQs and they may look nothing like today's fuzzy cameras and razors and washing machines. Some will have far higher machine IQs than we have meat IQs. Small fast computers will permeate our lives and work and play. And large systems and networks will get smart too. Linked satellite communications. Linked entertainment and gambling. Linked health and credit nets. Linked car and street and traffic nets. Linked smart weapons and spy nets. Linked news and poll nets. Linked government nets that watch, tax, measure, fine, credit, debit, and correct us.

In this last part I do not want to talk about future smart gadgets or systems or how they work. That talk belongs in journal papers and sci-fi stories. And a lot of the how part I have already talked about. Machines will get FATter. They will shrink and have finer sensors and signal processors. They will grow their own fuzzy rules with neural nets. They will paint pictures of the world with fuzzy cognitive maps. They will do it on a small but vast and fast scale. Vast and fast.

I want to talk about how the fuzzy principle may change our minds. I want to speculate on how fuzzy logic may change our views of life and death, our personal ethics and our systems of law and state, and our views on why we are here and what a God might look like if we find One.

Scientists are not supposed to talk about these things. And most philosophers no longer talk about them. Such has logical positivism won the day in the world of mind. All statements are true or false. Science finds out which. You can't ask "why" questions. If you can't test it with data or prove it with math, you can't say it. That was the whereof that Wittgenstein spoke of and said thereof we could not speak. But I will speak of it anyway just like in old-time philosophy.

Philosophy was different in the old days. You said all you could say with science and math and then you tried to see and say more. There just was not much science or math in the days of Descartes and Locke and Kant. Today there is and I have spent years in it and adding a little bit or fit to it and have gotten a feel for where the frontiers lie.

I want now to speculate at the edge of science. I want to go beyond the edge as a ladder goes beyond the firm ground that supports it. I will start where life starts and stop where it may end.

13

Life and Death

Do you see this egg? With this egg you can overthrow all the schools of theology, all the churches of the earth.

Denis Diderot

The most savage controversies are those about matters as to which there is no good evidence either way.

Bertrand Russell

I have looked over the wall and seen the bodies floating in the river, and that will be my lot also.

Unknown Sumerian or Babylonian
The Epic of Gilgamesh

They said to themselves in their deluded way: "Our life is short and full of trouble, and when a person comes to the end there is no remedy. No one has been known to return from the grave. By mere chance were we born, and afterwards we shall be as though we had never existed, for the breath in our nostrils is but a wisp of smoke. Our reason is a mere spark kept alive by the beating of our hearts, and when that goes out, our body will turn to ashes and the breath of our life will disperse like empty air. With the passing of time our names will be forgotten, and no one will remember anything we did. Our life will vanish like the last vestige of a cloud. And as a mist is chased away by the sun's rays and overborne by its heat, so too will life be dispersed. A fleeting shadow—such is our life, and there is no postponement of our end. Man's fate is sealed: no one returns."

THE WISDOM OF SOLOMON
THE APOCRYPHA

The principles of physics, as far as I can see, do not speak against the possibility of maneuvering things atom by atom.

RICHARD P. FEYNMAN
"THERE'S PLENTY OF ROOM AT THE BOTTOM"
MINIATURIZATION

Some people want to achieve immortality through their works or their descendants. I want to achieve immortality through not dying.

WOODY ALLEN

Does life start at conception? Do you murder a fetus if you abort it? Where do you draw the line between life and death?

Life starts at conception all right. It starts there because cell growth starts there. But the question is to what degree. That is what fuzzy logic adds to the debate. Degrees.

The term *life* is fuzzy. It shades smoothly into *not-life* or *death*. The shading is a matter of degree. It starts with how we learn the terms *life* and *living*. We learn them from example. Our neural nets learn them from example. So we learn to recognize life and the living. We don't learn how to define them. Recognition precedes definition.

Parents and teachers and books and TVs point out living and dead viruses, cells, plants, trees, fishes, animals, and humans. We see people young and old and in between and see them newborn and stillborn. We try to label everything as living or dead. As kids we first call life what moves. Life as animation. TV and cartoon characters move and live and populate our world. As we grow we draw new lines between life and death. We come to see life as—as what? A brain state? A set of physical behaviors? A DNA-like skill to replicate self? A bundle of legal rights? The borders blur but still we round them off and call it all living or dead. We have to call each thing alive or dead. The world has conditioned our minds that way. The bivalent reflex. The either-or reflex. This reflex lies at the heart of the abortion debate.

The abortion debate comes down to a black-and-white view of life versus a gray view. Or it will come down to it. The signs and sound bites for "pro-life" and "pro-choice" don't mention a fight between bivalence and multivalence. Maybe someday they will. Now the debate deals with women's rights and fetal rights and facts in the lab and clinic. I want to stick with the logic of the debate and that starts with the logic of lines.

LIFE LINES

Fuzzy sets throw a new set of tools into the fray. Sets and fuzzy sets can model the many views in the debate. But if you think life starts at conception *and* starts there 100%, you don't need a fuzzy set. A black-and-white set will do (Figure 13.1).

The pro-life view draws a line between death and life. It splits the universe into a life piece and a death piece. Death falls to the left of conception and life to the right of it.

The pro-life view seems well defined. It boils down to the old saw that you can't be a little bit pregnant. Mom and dad each

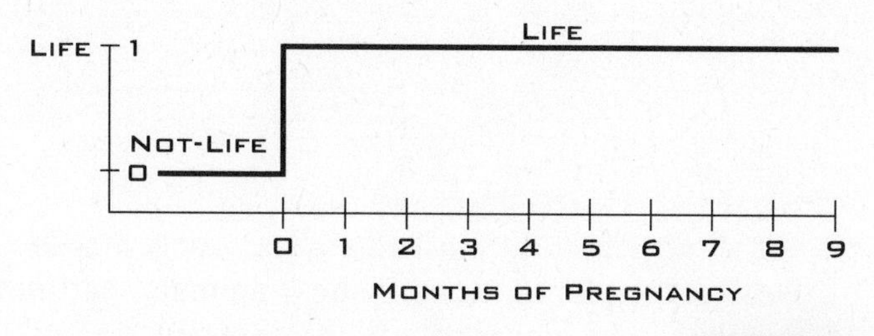

FIGURE 13.1

give 23 chromosomes. The sperm cell or needle puts dad's 23 in the egg with mom's 23. Life starts when, and only when, the 46 chromosomes fuse into a zygote. The fusion is a slow and sloppy molecular process but we round it off to a crisp event. Then the zygote lives if its cells divide in the right ways. Growth has started. Motion has started. Animation.

If you now define life as growth, then life has started too. But that is name calling, not argument. We can just as well call growth *death* and say that death begins at conception. That would use the term *death* in a new way and play off of the old way. We could always play with words and find a few examples or slogans from rare cults to make a case for the death-as-growth claim. To define is to label is to name-call is to do nothing at all. Who says growth is life? Who says it is 100% life? To draw a line is not to defend the line. It just gives us one case to look at. You can draw the line anywhere to the right of conception. The pro-choice view does just that.

The pro-choice view just shifts the pro-life line. It shifts it to the right. It draws a hard line to the right of conception but where does it draw it? Death still lies to the left of the line and life to the right. But where does the line fall? The 1973 *Roe v. Wade* Supreme Court case helped firm up the view that you should not abort a fetus past the first trimester. So you should draw the line at three months (Figure 13.2). Some pro-choice fans do not want abortions in the third trimester. They draw the line at six months (Figure 13.3). And some pro-choice fans say you can abort until birth. They draw the lifeline at nine months (Figure 13.4).

Pro-life fans denounce these many lines of pro-choice fans as arbitrary and unprincipled. The far-left line is well defined and

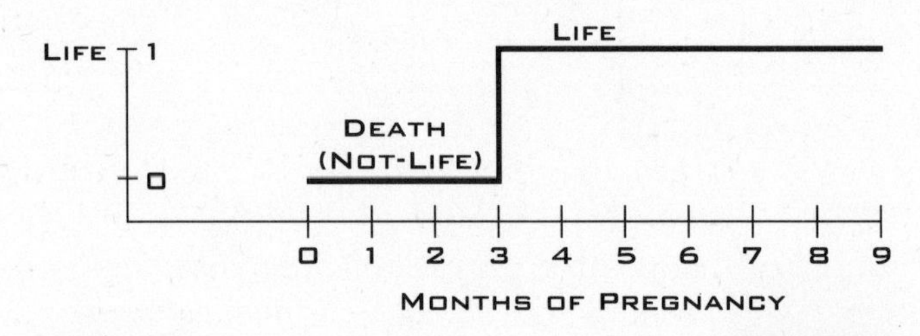

FIGURE 13.2

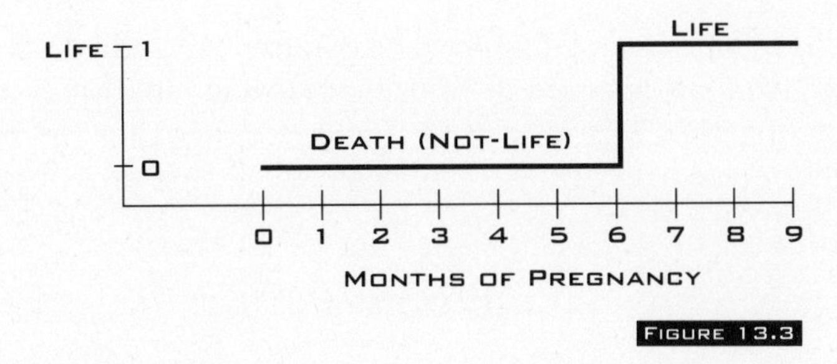

FIGURE 13.3

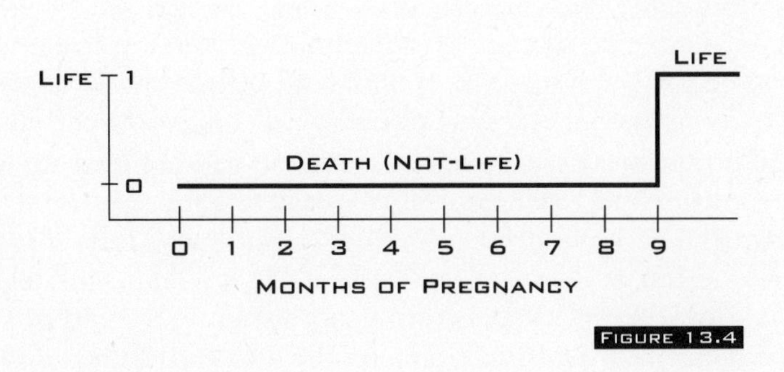

FIGURE 13.4

tied to cell growth. But the far-right line is also well defined and tied to birth. Between these extremes there are as many lines as there are numbers on the number line. So the abortion debate comes down to two debates. First the big one: the far-left line versus all other lines—pro-life versus all else. Then the small one: all other lines versus all other lines—the pro-choice spectrum.

And don't count on medicine and brain waves to draw the line. They only fuzz it up. An EEG measures smooth changes in brain activity. It is precise but that does not help. Precision increases fuzziness. You can't even see the line between cell and not-cell if you look at a cell with a big enough microscope. Zoom in and fuzz up. So the trick of defining death as brain death and then defining life as not-brain-death won't work. An EEG gives a curve. Anywhere you draw the line through the curve you get borderline cases. No one wants to kill or prosecute for want of a few electrical impulses. But EEGs do help since they remind us that death is gray. Some humans end up in comas and their family, friends, physicians, and insurers have to think about where to draw the death line. If death is gray and if death is not-life, then life is gray too. Life is fuzzy. You can draw lines through fuzz just as when in 1494 Pope Alexander VI drew a line through the New World to split it into one piece for Spain and the other piece for Portugal. The lines will be just as arbitrary.

LIFE CURVES

The problem is that lines are bivalent. Everyone plays with Aristotle's rules. No one can draw a line and get the two sides to agree. Each line is arbitrary. Each line is a fake split between life and death. Each line brings with it all the problems bivalence has caused in the last three thousand or so years. Forced round-off and borderline cases. There is a reason why we give students an A− or C+ or D instead of just a pass or a fail.

So let the Buddha draw a line and a not-line. *Draw a curve.* Life is a matter of degree. Death or not-life is a matter of degree. We all know that life shades into death when we are old or sick. In the same way death can shade into life. But the shades are gray. Life does not jump from off to on or from on to off just as history does not jump at the end of a century. Life unfolds from death to death.

A fuzzy life curve increases from left to right in Figure 13.5. The curve does not have borderline cases as a line has. To some degree no case is borderline. And to some degree every case is borderline. There is no abrupt change from death to life. A life curve gives a death curve and the two curves cross at the "midpoint" or 50% value in Figure 13.6.

Fuzzy curves have two big problems. First, there are infinitely many curves. There are as many curves as there are lines. Each person can draw her own curve and no two curves may ever be the same. Second, what do you do with a curve? We know what to do with a line. Punish those who go beyond it. We can argue over how to punish and how to be sure who went beyond the line. But in theory a line decides the case for us. What does a curve do?

It lets you nod your head. That's the first and best thing a curve does. It helps us agree. For the moment don't count the side issues. Just look at two cases on a ballot or on a news call-in show. Case one draws a line to split life and death. Case two draws a curve. Which would you vote for? It depends. No one

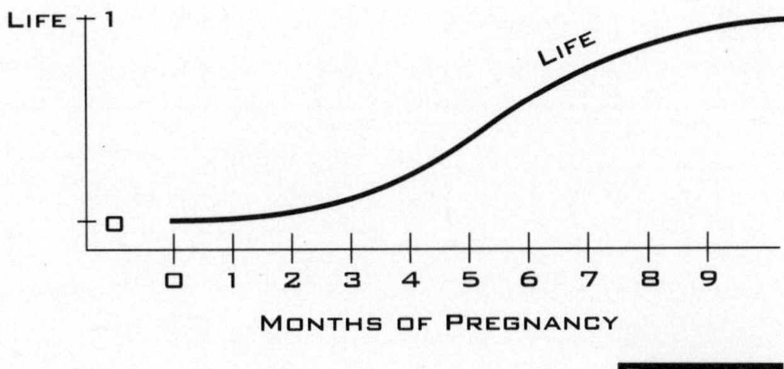

FIGURE 13.5

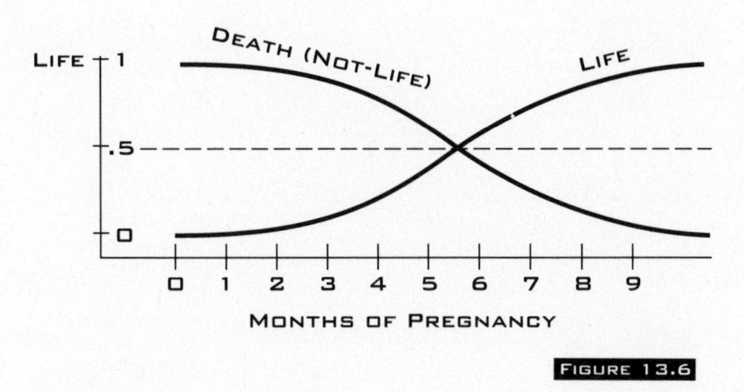

FIGURE 13.6

will vote for a bad line or a bad curve. So say the line and curve are close. The curve rises up sharply near where the line falls. Then I think most will vote for the curve. It gives a hedge. It smears out and softens the decision. But the first task is to find a fuzzy life curve.

TAKE A POLL AND FIND OUT

We can grow a life curve from polls. There are many ways to come up with fuzzy set curves from data. They all use statistics. They grow the curve as a staircase that rises from left to right.

A pollster might ask you if you accept abortions in the first month of pregnancy. If you say no, the pollster writes down a 0 for the first month and writes 0s for the next eight months. If you say yes, she writes a 1 for the first month. She then asks you if you accept abortions in the second month. If you say no, she writes a 0 for the second month and writes 0s for the next seven months. If you say yes, she writes a 1 for the second month (and has already written a 1 for the first month) and asks you if you accept abortions in the third month and so on. In the end you give her a list of 0s and 1s and maybe some borderline answers that she writes down at 1/2. She then adds up all the 0s and 1s in each month and divides the sum by the number of persons polled. This gives a nine-step curve that approximates a smooth curve (Figure 13.7).

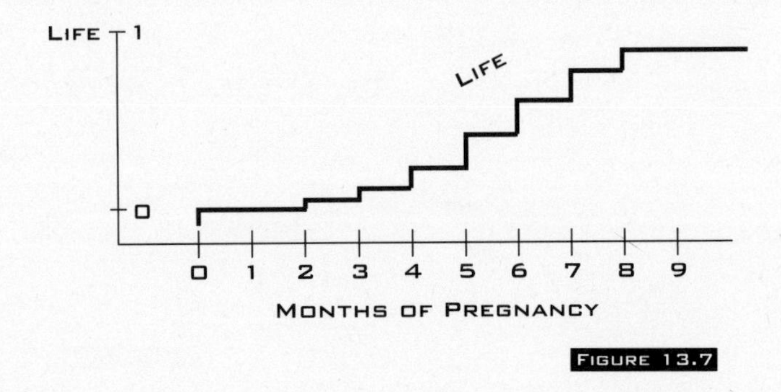

FIGURE 13.7

You get the same thing if the talk show gives you a phone number to call and you tell the show the *last* month you will accept an abortion. There are many variations on this but they all come down to averaging poll results. The averages get more and more accurate as more people take the poll.

This talk of polls may offend you. To some degree it offends me. I have never seen a pollster. How can you trust a decision you did not make or help make? I have studied statistics and know that the math is right and that it works well if you sample thousands of cases, whether thousands of light bulbs or front tires or persons. But polls still seem to dehumanize us. They bring out the mob side of democracy. You can cheat with how you ask the poll questions. You can cheat with how you describe the poll method. You can get the poll samples wrong even if you do your best to get them right. You can skew polls with false or slanted press releases or with shock radio or shock talk shows. And the press and politicians play with polls in the name of science. The press has made an art of the science of polls. The press acts as if it has answered the oldest question of ethics: What is right? Take a poll and find out.

And that's just what I suggest we do. Take a poll and find out. Take lots of polls. Take lots of polls in lots of states and professions and countries. Would you not like to see the shape of the U.S. life curve? It is out there just as the average height of U.S. males at ages 5 and 10 and 15 and 20 is out there. It floats in our minds and brains as a statistical absolute. All we have to do is fish it out. So take a poll and find out.

In time each state or country may publish its fuzzy life curve. Medical groups might put forth their own. Special panels and think tanks may average the many life curves and weight them with population numbers. Each year or so the American Medical Association could send a poll or draw-your-curve chart to their members.

Life curves are not pro-life or pro-choice. Most curves do not support an extreme pro-life stance. But besides that they just help us make more informed decisions. That fits well with a diverse people. Few persons in the U.S. take an extreme stance on abortion.*

A life curve gives more information than a life line. It paints a more accurate picture. It helps us agree on *something*. But it does not do what we really want it to do. It does not give us a line. It does not slake our thirst for either-or action. A life curve offers only a limited advance. But the abortion debate has not advanced since 1972 and then only in law terms.

Now suppose we know the U.S. life curve. At any time it is a "real" and fixed statistical thing. The best we can do is sneak up on it with lots of polls. So suppose we have. Suppose the U.S. life curve is so familiar that medical groups and high-school science and civics classes and the Supreme Court cite it when they discuss abortion. What would happen?

I think we would take the life curve and draw life lines through it. These lines would not be the old pro-life and pro-choice lines.

---

*Steven Rosenstone and Maria Calvo of the University of Michigan, in "Public Views on Abortion Show Little Consistency," *The Wall Street Journal*, April 22, 1991: "Only 17% [of pro-choice advocates] have completely 'pure' positions for abortion rights, the study concludes, and only 21% [of pro-life advocates] have completely 'pure' anti-abortion positions. Declare the authors: 'Few Americans have both feet in either camp.' "

They would fall in a window around the "hump" in the curve. Which hump? I don't know. It depends on the curve. Maybe the lines would cluster around the midpoint or 50% mark. These lines would give us the new debate. The lines on the left would define the new pro-life stance. The lines on the right would define the new pro-choice stance. Counselors and physicians would talk of this "window of life" to their patients. They would refer to it as a guideline. If we pass laws on abortion, they might well deal with this window. In any case we would use the new information in the published life curves. It would slowly change the way we think about abortion. It might even cool the debates.

DEATH IN DEGREES

Then there is death. If life has a fuzzy boundary, so does death. Most of us accept that. The medical definition of death changes a little each year. More information, more precision, more fuzz. Every day paramedics save men and women from heart attacks and other calamities. The heart stops and then it starts again. The brain waves fall flat and then they ripple again.

Fuzzy logic may help us in our fight against death. If you can kill a brain a cell at a time, you can bring it back to life a cell at a time just as you can fix a smashed car a part at a time. When does a cell die? There is no hard line. All our cells are living or dying to some degree.

Think of armies of smart machines in your bloodstream and tissues and bone marrow. This is the frontier of *nanotechnology* and the smart machines are tiny robots or *nanobots*.

Nano means billionth as in billionth of a meter. Molecular machines. Machines that build and fix parts a molecule at a time and that destroy parts a molecule at a time. Today our computers use *micro*processors that work on the scale of a millionth of a meter. They deal with big blobs of molecules and cannot stack or unstack them a molecule at a time. Microprocessor chips have hit their quantum limits. Circuit density on chips has doubled roughly every two years since the 1970s. To keep up the shrink we have to move into the nano world.

Today many researchers in the U.S. and Japan are designing nanoprocessors and polymer or plastic computers. In 1988 CalTech's John Hopfield, a pioneer in the field of neural nets, designed a polymer memory chip. He has not built it yet but he has worked out its design. Pulses of light both synchronize the chip's "clock" and give it energy. The polymer is a long repeating chain of molecules. The polymer chip puts things in memory and takes them out by passing chemical messages down the polymer chain as firemen used to pass buckets of water in a fire. Other nanochips may work with locks and keys and molecular gear systems.

Now think of armies of nanobots that pass messages and reason with tiny fuzzy brains and breed like viruses when they have to. In theory these nanobots can kill AIDS and cancer and disease cells and can repair and boost old cells and tissue and might even sharpen our eyes and heighten our taste buds and clean our teeth. Swarms of nanobots in the air might eat pollutants and acid rain and keep the ozone layer strong. In water they might eat poisons and clean up Lake Erie. This won't happen in a decade but it will happen. It will come to pass. And there is a reason for this.

Death is an engineering problem. Death of the planet. Death of ecosystems. Death of the body. There are no ghosts or spirits or souls at the nano level and that is a good thing. Ghosts and souls do not obey the laws of physics. Molecules do. The system dies when its parts die. The part dies when its cells die. The human body has trillions of cells. Each cell has more than tens of thousands of molecules in it. The molecules fit together in patterns. The cells fit together in patterns and the DNA code tells how they fit. We can rebuild these patterns a molecule and a cell at a time and in time improve them. Molecular engineering. Molecular machine intelligence. Fuzzy logic in a nanobot. Man a fuzzy machine.

Molecular engineering breaks no laws of physics or chemistry. As Nobel physicist Richard Feynman said in the early 1960s, there is plenty of room at the bottom. He even offered a prize: "I hereby offer a prize of $1,000 to the first guy who can take the information on the page of a book and put it on an area 1/25,000 smaller in linear scale in such a manner that it can be read by an electron microscope." Most scientists dismissed that challenge as cute but empty or too far off to count. Then in 1990 Don Eigler at the IBM labs used an electron microscope to drag and stack 35 xenon atoms to spell out IBM. *Science*, *Nature*, and most newspapers ran the picture on their covers or front pages. Spell with atoms. Man's first use of atoms to encode information. The first assault on the tyranny of matter. Why not turn a statue into a man? Eighteenth-century philosopher Denis Diderot found a way to do it. Grind up the statue into powder. Use the powder to fertilize beans and peas and other green things. Then eat the green things. Molecular engineering suggests a simpler approach: unstack the statue molecules and restack them as man molecules. Along the way you bring the dead to life.

Most of us will not live to see the age of fast and cheap molecular engineering. The first generation that does may live for a long time. Future generations, if they still have generations, may not die by disease or decay. Accidents may kill them and then the smart swarms of nanobots may fix them. Or they may die by choice. By then it will be legal.

For the rest of us there is one known way to time-travel to that future. It is not a good way but right now it is the only way. We can freeze our bodies or just our brains at very low temperatures. We can do time in a bottle of liquid nitrogen at about −320° Fahrenheit. That is cold enough for most molecular change to stop. There is some quantum decay. And the extreme cold so stresses flesh that it cracks organs and the brain in big and small fissures. Thaw out and you look worse than a thawed and soggy strawberry. But in theory smart nanoengineering can fix the strawberry a molecule at a time. And in theory it can fix you the same way. You might have to wait a few centuries to get fixed. In that time society might break down or the guys in white smocks might forget to refill the liquid nitrogen that boiled away or the state may forbid cryonic revivals or the guys in white smocks might want to keep those trust funds that have grown exponentially. Even if it works you might wake in pain or in an idiot haze or as a state slave or as a living-parts shop. Or it may turn out well and you come out 18 again and get to water a lot of old graves.

I don't think it would have hurt to freeze Einstein or Gandhi or a few dozen others in the twentieth century. We had the glycerin to embalm them and the liquid nitrogen to freeze them. Einstein gave his brain to science the old-fashioned way and it got pickled and then lost. Physicists in a hundred or thousand years might curse us for ruining Einstein's brain. They might say, as the cryonicists now say, that a mind is a terrible thing to waste, bury, or burn.

Or look at the eco-systems we destroy. Every day we lose a few more species as we "develop" the planet. Why not freeze a rare moss or fern or herb or tree or freeze a mating pair of rare fish or monkeys or birds or insects? Put them in a cryonics ark. The United Nations or environmental groups might tend it.

And why not freeze some of those aborted fetuses? Most abortion methods destroy the fetus. But in time that may pass. Very early abortions can preserve the embryo. We already freeze or "vitrify" embryos and store them and then thaw them and put them in new wombs where they grow the way you and I did. Frozen embryos might suspend the abortion debate. Churches or the Vatican could store the embryos in their basements and work out plans to thaw them and raise them. They may never thaw them all but it might make pro-life fans feel better.

The rest of us who are not Einsteins or rare species can pay our money and take our chances. As of 1992 there were only about 300 persons with cryonics contracts and only about 30 in the cryo-freezer. Most in the tank float there as heads only. A tank of heads rather than bodies is cheaper and easier to tend and to protect from earthquakes and from the organized morticians and the county coroners who oppose cryonics.

I am one of the 300 who have signed up. I went for whole-body suspension. I like the idea of taking those extra parts with me. My friends dismiss this as an expensive burial. Or they question whether if it works I will have any memory that keeps me me. If there are big memory gaps, maybe I can fill them in part by reading old newspapers and watching old TV shows and movies or by reading my old papers and books. I don't know. I want to see the fuzzy future and I don't know how else to do it. I would not bet my life on cryonics. But I am happy to bet my death on it.

14

Ethics and the Social Contract

Man is the measure of all things.

Protagoras

"Listen then," Thrasymachus said, "I say that justice or right is simply what is in the interest of the stronger party."

Plato
The Republic

There is no such thing as justice in the abstract. It is merely a compact between men in their various relations with each other, that they will neither injure nor be injured.

Epicurus
Principal Doctrines

When a covenant [contract] is made, then to break it is unjust. And the definition of injustice is no other than the not performance of covenant. And whatsoever is not unjust, is just.

Thomas Hobbes
Leviathan

Man was born free, and everywhere he is in chains. When the State is established, consent lies in residence. To dwell in the territory is to submit to the sovereignty.

JEAN-JACQUES ROUSSEAU
THE SOCIAL CONTRACT

Science, thanks to its links with observation, retains some title to a correspondence theory of truth. But a coherence theory is evidently the lot of ethics.

WILLARD VON ORMAN QUINE
THEORIES AND THINGS

There are no objective values.

J. L. MACKIE
ETHICS

Science is the measure of all things. It measures the universe and its pieces with telescopes, microscopes, speedometers, spectrometers, barometers, electrocardiograms, breathalizers.

And we measure ideas against science. Science favors a lean diet and exercise and washing your hands. It does not favor ideas that don't fit with natural selection or the Big Bang or the second law of thermodynamics. Science has all but ended astrology and philosophy and religion. They live on as shells of what they once were.

So what about ethics? Does science dispose of morals and law and the social contract that we live by?

I think the answer is fuzzy: yes and no. In some sense reason always ends in doubt. Each day more facts pour in and our measurements of the world get more precise. So issues blur and our view gets fuzzier. Every thing causes every other thing to some degree. Tobacco smoke and burnt food and sunshine and asbestos and TV sets cause cancer to some degree. A million forces in the economy help warm the planet and eat its thin atmosphere. A million forces in our brains and bodies and genes shape our thoughts and choices and behavior. Each day it gets harder to call one thing or event or act good or bad or call it right or wrong. The lines blur into curves. Reason ends in doubt. What starts out black and white ends up gray.

Fuzzy logic is a part of science. It came late but it came. So what does it say about ethics? You may not like it. In a sense that I will explain science has disposed of ethics. We have no final argument but force against the young men who run through Dostoyevsky novels and shout "Everything is permitted!" On top of this, fuzzy logic blurs the ideas of ethics. Each act is to some degree intentional and determined. Each law admits degrees and a spectrum of exceptions. Up close the social contract shades into gray and frays into whatever you want it to be.

Science undercuts ethics because we have made science the measure of all things. Truth means truth of science. Truth means logical truth or factual truth. Truth means math proof or data test. The truth can be a matter of degree. But that does not help ethics.

THE FUZZY TRUTH OF ETHICS

The argument starts like this. Are ethical statements true? Are they false? Look at the ethical statement "Murder is wrong." Is it true? We know that the sociological statement "Most people believe murder is wrong" is true. But what about the pure moral claim? Is murder wrong? We want to say yes and we do say yes and we back it up with force. But does that mean murder is wrong?

Scientists look for truth in math and tests, in logic and fact. So start with logic. Is "Murder is wrong" logically true? If so, it is true by definition or by derivation. Is "Murder is wrong" true by definition? If so, it is trivially true. We have just defined it that way. We can define its opposite the same way. Truth by definition is not truth at all. It is name calling.

Can we derive "Murder is wrong" as a theorem as we derive the Pythagorean theorem? We can derive it from other moral claims like "Unprovoked aggression is wrong" and "Murder is unprovoked aggression." But that just backs up the argument to the first moral claim "Unprovoked aggression is wrong."

We want to derive *ought* from *is*. Philosophers have tried to do that for thousands of years. No one has derived it. You get no more out of an argument than you put into it. You assume premises and derive conclusions. If you assume an *ought* in a premise, you can derive an *ought* in a conclusion. But that begs the question and just moves the debate up to the level of the premise.

This is half of what Quine meant when he said that "a coherence theory [of truth] is the lot of ethics." You can assume a moral axiom like the principle of nonaggression—whoever first uses force, or threatens to use it, is wrong—and then grow a moral tree from that root. Logic takes you from the tree root to a branch to a leaf. The moral sap flows up the tree. The moral statements cohere. They do not contradict one another and each comes from others except the axiom root, and that comes from assumption. To cohere is just to be "internally consistent." A fairy tale coheres.

THE TEST OF MORALS

The other half of Quine's claim deals with fact and whether our ethics match facts. We want our moral tree or its root or at least some branch or leaf to be true in fact. We want our moral claims to correspond to fact. We want our moral claims to have more truth than fairy tales. Logical truth is empty. In the end it is just symbol pushing. Here again we face the mismatch of logic with gray world, with fact. The real power comes from a statement that matches fact. Fairy tales don't match fact.

So is it a fact that murder is wrong? We know it is a fact that we believe murder is wrong. Or do we? This claim is weaker than it looks. It holds only to some degree. Each term in the statement is fuzzy. We know things to some degree. We believe things to other degrees. The border of murder and nonmurder is not sharp, as any law student knows. What we mean by "wrong" is vague and fuzzy and hard to tell from "not wrong." This is how fuzzy logic helps make a mess of ethics and law. The paralysis of analysis. The best we can say is that to some high degree we believe that murder is wrong. This claim comes with a set of disclaimers and the hope that we do not have to fully define our terms. But grant the claim. Grant the fuzzy fact. After all, all facts are fuzzy. So we know at least that some statements *about* moral systems are true in fact to some degree. What about statements *within* a moral system? Is murder wrong?

How do you match a fact to a statement? All matches are fuzzy since all facts are fuzzy and all statements of fact are fuzzy. But again grant the match. Just don't forget you granted it. You get black-and-white logic that way.

You match a fact to a claim with a test. You test the claim. You hold an "experiment." You find data or "experience" that goes for or against the claim. You open the door to the world and let the flux of experience or events pour in. That flux is a chunk of the space-time continuum. So to match is to test. To test is to find a chunk of space-time.

But words and chunks differ in kind. The missing link in every claim of science is how we tie the words or math of the claim to the chunk of space-time that tests it. The words are fuzzy since each noun like "tree" or "atom" or "star" stands for a fuzzy set of trees or atoms or stars. Each verb like "flows" or "increases" or "suggests" or "changes state" stands for a fuzzy set of actions or states of being. Each term like "tends to" or "usually" or "exponentially" slants a verb to some vague degree. Each adjective like "stable" or "transient" or a predicate adjective like "is statistically relevant" defines a fuzzy subset of some set of noun things like "states" or "tests." Every day scientists make their favorite fuzzy claim that "The data *suggest* such and such" and they do not, and maybe cannot, explain this law of suggestion. And the chunks are fuzzy: the electron cloud, the molecular equilibrium, the lung cancer, the Spanish empire, the battle of Waterloo, the Great Depression, the supply of wheat, the demand for wheat, the Arctic current, the trade wind, the earthquake, the stratosphere, the ozone layer, the sunspot, the solar wind, the ring of Saturn, the Milky Way galaxy, the known universe. To tie words to chunks is to tie fuzzy sets to fuzzy blobs.

A claim is true if its words match fact. "Grass is green" is true if and only if grass *is* green. Grass is a fuzzy set. The set of green things is a fuzzy set. Green grass is a fuzzy blob of space-time. You test "Grass is green" when you tie the grass sets and the green sets to the blob of green grass. If the words tie to blobs, the claim is true to some degree. The tighter the tie, the truer the claim. How tight is a tie? No one can say. We nod at some ties and shake our heads at others and then move on to the next issue.

A truth test needs a chunk of space-time. But which space-time? We want to say "The one we live in" but we can't be sure that it is just one. We can't be sure of anything. For us being and being sure are states of neural nets in our brains. That comes down to electricity and chemicals and hormones and other brain stuffs. The only thing we can be sure of is logical tautologies like "Grass is grass" or "2 + 2 = 4" and those, as Einstein saw, do not apply to the real world. We may live in more than one space-time continuum. The new physics says tiny wormholes may form on the tip of your finger each second (in $10^{-43\text{rd}}$ of a second) and pop out in a new Big Bang in a new space-time. We can also picture how we could mix up our space-time and not break those few laws of physics we know. These are *physically* possible worlds. In some of them "Grass is blue" may match well with the blob of blue grass. "Grass is blue" is false, at least to a high degree, in our space-time. It may be true in others. So it is a *testable* claim. That means we can imagine some possible chunk of space-time to tie the claim to. We may never do it, just as we may never have a direct test of how the Big Bang blew up 15 billion years or so ago or how wormholes wiggle or how our universe may end in a gravity collapse or heat death.

In sum a factual claim is a testable claim. It tests true to some degree if it matches the right chunk of our space-time. It tests false to some degree if it does not match the right chunk. A claim is testable if it "could" match (if it matches) some chunk in some space-time. It is not testable if it "could not" match (if it does not match) any chunk in any space-time.

So can we test "Murder is wrong"? How? With what chunk? What would count as evidence for it? We think we know the fuzzy set of murder acts and the fuzzy set of wrong acts or wrong things. But what do we tie the string to? "Murder is wrong" is true if and only if murder *is* wrong. We can find chunks of space-time that are murder acts. But where do we find chunks that are wrong, are full of wrongness? No fact or data or observation or experience or measurement seems to make the claim true.

The same goes for whether "Murder is wrong" is false. What data would knock down the claim? What would falsify it? Every claim of science must risk falsehood. The claim must stand to fall if the facts don't turn out right, if the prediction fails. Nothing wagered nothing won in test. The more a claim risks in test, the deeper the truth if it survives the test. "Red is red" risks nothing in test. The equation "$e = mc^2$" risks it all in tests and so far passes the tests. But in the next test that may change. Moral claims seem to be neither logical truisms nor testable claims of fact. They seem untestable in principle. And they are fuzzy to boot.

This claim about morals is testable. To refute it you need find but one true *or* false moral statement. So far no one has. Think about that. Thousands of years of human culture have passed and no one has found a single moral truth or falsehood. Men have sought to ground their morals in God or philosophy or society or government, in the force of logic or fear or police—in force. But no one has found one shred of evidence for or against a moral hypothesis. In this sense science disposes of ethics. Truth here is not a matter of degree since it does not apply. So science disposes of fuzzy ethics as well.

Science can tell us what we think is right but not what is right. Science, as sociologist Max Weber said, acts like a road map. It can tell you how to get from one place to the next but not where to go. At best it says what is. We say what ought to be and we say it alone. We are free to tell right from wrong all right. What I tell you is right you may tell me is wrong.

RIGHT AND WRONG: HURRAH AND BOO

At this point you might ask what these moral claims are we make. We make them as if they are true claims and we back them up with force. What is left if we subtract out the force? What are ethical claims if they are not true or false statements?

Modern philosophy gives one answer: They are boos and hurrahs. They exclaim or exhort. A moral claim may express how we feel, as when we say "You should be punished" or "Greed is good" or "Greed is bad."Here we exclaim. We just write our autobiographies out loud. Or we exhort. We say or sing or shout moral claims to stir up emotion in our audience, to change how they will act in the future, to change their probability of action, as when we say "Theft is wrong" or "You should not lie" or "(You should) vote for us." Exclaim or exhort. Boos and hurrahs. Some moral claims both exclaim and exhort. They both express feeling and reinforce future behavior: "You did the right thing" or "You sinned." We *command* as we *commend*. The two differ by only a vowel.

Three things skew moral claims. First, they are fuzzy to the core. We all mean different things by "lie" or "fair" or "right" and to some degree every act is right and wrong, fair and unfair, just and unjust. Second, we have a stake in lying and hiding in the fuzz when the moral claim applies to us. Gain and loss. Wealth and property and liberty. We backpedal, fudge, fib, white lie, exaggerate, lie outright, trim and edit memories, and slant descriptions to help us and ours. And there are lawyers. Third, our deepest emotions side with our moral claims. A film or speech or book "moves" us by moving our hypothalamus and limbic system. What feels right in ethics has a long genetic history. What feels wrong has at least as long a history in the gene lines—how do you feel about incest with a condom? This gives our moral claims an intuitive sense of self-evidence and a license for self-deception. But our emotions arbitrate nothing. They just show that our endocrine systems work.

Our endocrine systems did not evolve in accord with right conduct. They evolved under pressure from the competitions for mates and scarce resources. Hundreds of millions of years of vertebrate evolution have shaped our endocrine systems, our mix of hormones and nerves. Hormones splash into, and speed up or slow down, most of the neural nets in our brain. Emotion and cognition are woven together in our meat computer. We share much of our emotional blueprint with mammals and reptiles and fish. They are our cousins in feeling. But we can put our feelings in words and symbols. That's where the fuzz comes in, the cognitive step, the thicker neocortex. If blackbirds or dolphins or house dogs could talk, and one day with smart genes and brain implants they may talk, their speech might be shot through with ethical exclamations and exhortations. For now a dog talks with barks and growls. Growls shade into barks and so dog talk is fuzzy. But who would say that a growl is true or false?

THE FORCE OF LAW

Reason ends in doubt. Science disposes of ethics. It strips moral claims of logic and fact and reduces them to feeling in words. And yet we have to make moral claims and fight for them and once in a while go to war for them. We have to exercise our moral freedom whether we like it or not. Here arises the law and the "social contract" that the law helps write.

Society rests on the social contract. We weave the "fabric" of society on top of it. We make rules and pass laws and elect leaders to change small pieces of the social contract. But who can say just what the social contract is? Economists call it what we do as a group that we can't do alone. But that ball of fuzz does not pin it down. A lot rides on the social contract. If science does not ground ethics, then the weight of ethics falls on the social contract. So where is it? Who wrote it? Did it just evolve as the English language did or as money and the price mechanism did or as Common Law did? Who signed the social contract? How do you sign it? Are we stuck with a mere verbal contract?

We start with law. Law is the set of fuzzy moral claims that society or the state backs up with force. "Murder is wrong" has no factual truth but it has legal "truth" or validity. This legal truth is the coherence kind. We agree on some legal rules and principles like freedom of contract and presumption of innocence. In theory we pack the root or *Grundnorm* rules and principles into a state's constitution. Then the legal tree or bush grows from the root and again the legal sap of validity flows up from root to branch to leaf. A *supreme* court interprets the constitutional roots. Its authority flows to states and congresses and executive branches and to their branches and on out to mayors and policemen who hide in speed traps. In practice we work with more rules and principles than we have written down or voted on. They are all fuzzy and subject to debate. There is so much fuzz that no one lies or no one has to admit a lie if she has a good lawyer.

I don't want to dwell on gadgets but in the future we may have good cheap legal advice on a chip. We may carry law aids to court with us in our palmtop computers. Today fuzzy systems work with rules of common sense. Tomorrow they can work with rules of moral sense. I mean this besides having the best lawyers you can put in a box. Everyone will have those and they will tend to cancel out. I mean moral advice. We know moral truth is too much to ask. But advice we can ask for. Smart fuzzy systems or cognitive maps can store and reason with the wisdom of the ages. We might ask advice from the Buddha or Confucius or Socrates or Aristotle or Machiavelli or Immanuel Kant or the new Pope or the latest pop guru or TV evangelist. Reading the masters comes down to asking their advice. Why not put their wisdom in fuzzy rules and put that in your computer or dashboard or briefcase or ear? It won't clear up the fuzz of ethics and law though it might make us feel better about it. In fact lawyers and Buddhas on a chip would weave a more tangled legal web. Everything seems to push in that direction.

THE FUZZY LABYRINTH OF LAW

Law is a fuzzy labyrinth. A legal system is a pile of fuzzy rules and fuzzy principles. And it is dynamic. Every day judges and legislators add new rules and laws and delete or overturn old ones. It is not an art or science. You don't play the game unless you have a stake in it. Then you tend to lie and deceive and cheat at least to some degree. Here the fuzzy principle holds with more force than anywhere else in our lives. Everything is a matter of degree. Legal terms and borders are fuzzy. Try to draw a line between self-defense and not self-defense or between contract breach and not breach. The lines are curves and you have to redraw them in each new case. Every rule, principle, and contract has exceptions. Each day judges, lawyers, and juries find new exceptions. Each "fact" in a case melts when you look at it up close and put it through the fires of cross-examination. Lives and careers and psyches depend on how we split balls of legal fuzz into A AND not-A, how we work with guilt, intent, premeditation, malice, threat, duress, equity, fairness, reasonability, acquiescence, duty, obligation, partiality, conflict of interest, damage, and property right.

The labyrinth has hallways and has cousin labyrinths that it deals with from time to time. States write their own constitutions and penal codes. The same federal constitution justifies them in some big fuzzy sense but on small points they disagree. They disagree in what they say and how lawyers interpret what they say. In Oregon highway patrolmen can use radar to measure car speed and give speeding tickets. Next door in California they can't. Laws and punishments vary across states. Judges in each state, city, and courthouse hand down different verdicts and sentences in similar cases. Each city passes laws and rules and builds its own fuzzy labyrinth of law. Each state and each country builds bigger labyrinths. Each day new cases make the labyrinths spill over on one another and tangle and pile up and grow new hallways.

Science does not help. It digs more tunnels in the labyrinth. First it strips truth from ethics. Then it gives tools that make the small things more precise but the big things fuzzier. Forensics labs find tiny facts of blood type and dirt type and tobacco type and dinner type. Lie detectors and vocal stress analyzers measure smooth changes in body function. This just gives more fuzzy facts to interpret, to fit in a fuzzy theory. The increased precision increases the fuzz.

DNA fingerprinting can tell who was there but not what they did. Videotapes can show part of what they did but not why they did it. More facts open it up to more interpretations. A video can show that Mr. X shot Mr. Y or hit him with a car or fist or pushed him off a bridge. But a video cannot draw a line between self-defense and murder or between murder and accident. That depends on Mr. X's plans and goals and intentions. As philosopher H. P. Grice said, you mean or intend one thing if you show a husband a picture of his wife with another man. You may mean or intend something else if you *draw* him a picture of the same thing.

THE FUZZY SOCIAL CONTRACT

The whole legal system rests on the social contract. A fuzzy social contract binds us. It limits our freedoms as it helps protect them. The social contract lies deeper than the constitution and legal labyrinth that sit on top of it. No one sees it. No one writes it down. It combines fuzz with invisibility and so it can give birth to surprises. It does not give us the rule of law. The social contract gives us *a* rule of law. There are many.

Would you sign the social contract if you saw it and studied its fuzzy duties and fuzzy benefits and fuzzy costs? You and I did not sign the constitution but the state says we have consented to it. Consent lies in residence. Staying is playing. Love it or leave it. The state says nothing about the social contract. It comes before the state and it seems to be a verbal contract between parties unknown. So if you could see it, the question whether you would sign it is a valid one. Would you sign the social contract as is or mark it up? Would you try to squeeze out the fuzz and replace gray curves with black-and-white lines? Or would you leave the fuzz in to keep things flexible and so trust that reasonable men and women will do reasonable things?

Here's a social contract you can sign. Computer engineer Robert Alexander wrote down his version of our social contract. I think he got the fuzz about right. He titillated hordes of hackers when he sent it out, free of copyright, on several computer networks. If you sign it, you can send it to the White House.

SOCIAL CONTRACT*
between an individual and the United States Government

WHEREAS I wish to reside on the North American continent, and
WHEREAS the United States Government controls the area of the continent on which I wish to reside, and
WHEREAS tacit or implied contracts are vague and therefore unenforceable,

I agree to the following terms:

SECTION 1: I will surrender a percentage of my property to the Government. The actual percentage will be determined by the Government and will be subject to change at any time. The amount to be surrendered may be based on my income, the value of my property, the value of my purchases, or any other criteria the Government chooses. To aid the Government in determining the percentage, I will apply for a Government identification number that I will use in all my major financial transactions.

SECTION 2: Should the Government demand it, I will surrender my liberty for a period of time determined by the Government and typically no shorter than two years. During that time, I will serve the Government in any way it chooses, including military service in which I may be called upon to sacrifice my life.

*Copyright 1989 by Robert E. Alexander, *Liberty*, 1989. May be copied and distributed freely.

SECTION 3: I will limit my behavior as demanded by the Government. I will consume only those drugs permitted by the Government. I will limit my sexual activities to those permitted by the Government. I will forsake religious beliefs that conflict with the Government's determination of propriety. More limits may be imposed at any time.

SECTION 4: In consideration for the above, the Government will permit me to find employment, subject to limits that will be determined by the Government. These limits may restrict my choice of career or the wages I may accept.

SECTION 5: The Government will permit me to reside in the area of North America that it controls. Also, the Government will permit me to speak freely, subject to limits determined by the Government's Congress and Supreme Court.

SECTION 6: The Government will attempt to protect my life and my claim to the property it has allowed me to keep. I agree not to hold the Government liable if it fails to protect me or my property.

SECTION 7: The Government will offer various services to me. The nature and extent of these services will be determined by the Government and are subject to change at any time.

SECTION 8: The Government will determine whether I may vote for certain Government officials. The influence of my vote will vary inversely with the number of voters, and I understand that it typically will be minuscule. I agree not to hold any elected Government officials liable for acting against my best interests or for breaking promises, even if those promises motivated me to vote for them.

SECTION 9: I agree that the Government may hold me fully liable if I fail to abide by the above terms. In that event, the Government may confiscate any property that I have not previously surrendered to it, and may imprison me for a period of time to be determined by the Government. I also agree that the Government may alter the terms of this contract at any time.

____
SIGNATURE

____
DATE

15

MAN AND GOD

God eternally geometrizes.

PLATO

What we observe is not nature itself, but nature exposed to our method of questioning.

WERNER HEISENBERG
PHYSICS AND PHILOSOPHY

A slight variation in the axioms at the foundation of a theory can result in huge changes at the frontier.

STANLEY P. GUDDER
QUANTUM PROBABILITY

Emptiness is like water. Existence is like its waves.

HSING YUN
LECTURES ON THREE BUDDHIST SUTRAS

I was not looking now at an unusual flower arrangement. I was seeing what Adam had seen on the morning of the creation—the miracle, moment by moment, of naked existence.

ALDOUS HUXLEY
THE DOORS OF PERCEPTION

The lies we tell about our duty and our purposes, the meaningless words of science and philosophy, are walls that topple before a bewildered little "why."

JOHN STEINBECK
THE LOG FROM THE SEA OF CORTEZ

Call no man happy before his death.

SOLON

Life is suffering.

THE BUDDHA

The utter vulgarity of the herd of men comes out in their preference for the sort of life a cow leads.

ARISTOTLE
ETHICS

The no-mind not-thinks no-thoughts about no-things.

THE BUDDHA

The human body is a machine that winds its own springs. Everything depends on the way our machine runs.

JULIEN DE LA METTRIE
MAN A MACHINE

The real question is not whether machines think, but whether men do.

B. F. SKINNER

Civilization merely develops man's capacity for a greater variety of sensations, and absolutely nothing else.

FYODOR DOSTOYEVSKY
NOTES FROM UNDERGROUND

We say to ourselves: it would indeed be very nice if there were a God, who was both creator of the world and a benevolent providence, if there were a moral order and a future life, but at the same time it is very odd that this is all just as we should wish it ourselves.

SIGMUND FREUD
THE FUTURE OF AN ILLUSION

When the Old God goes, they pray to flies and bottle tops. The future belongs to crowds.

DON DELILLO
MAO II

Why is there something rather than nothing? Why stuff rather than no stuff? Why not just void?

That is the oldest question of philosophy. It is a question of first philosophy. Metaphysics. Ontology. It is the grandest why question of them all. And modern philosophy says you can't ask it. It is a "pseudo-question." It looks and sounds like a real question. But it is not, the claim goes, since you do not know what would count as an answer to it. That scares them off. If you can't lay out a set of "meaningful" answers, you can't ask.

I do not accept a hard line between what counts as an answer and what does not. The same goes for "meaningful" versus "not meaningful." Hard lines are for the fictions of math. If you have a question, ask it. Steinbeck saw that. Sooner or later the audience sneaks in a "Why?" and the dam cracks.

I want to know why there is something rather than nothing. Maybe someday aliens from the Crab Nebula or one of their robot scouts with an IQ over 1,000 will tell us. Maybe we can find some answers that make sense. In this final chapter I want to start with this forbidden question and give a fuzzy answer to it. You may not like my answer. But it is an answer and there are not many answers to that question outside of religion. Then I want to turn to the heart of religion, to God, and paint a picture that has come to me in my work with fuzzy logic and neural nets. Again you may not like the picture but it can't hurt to see it. Last I want to turn to the image of this God, to man, and look at how more machine intelligence will change man and his world. Fuzzy logic raises machine IQs just a little. Soon they will soar. Where will it end?

THE ANTHROPIC PRINCIPLE: It Is So That I Am

So why is there something rather than nothing?

The old answer of religion is that God made the world and so there is something. Then the regress starts. Is God something or nothing? If God is something or some thing, who made him? No answer puts you back to the first question. We could just as well say "X made the world" if we do not define X and tell where X comes from. If God is nothing, how could he make something from nothing? And if God is nothing, then God is just a fiction, just the vacuum, just the empty set—God is everywhere in the sense that he is nowhere.

Greek philosophers played with these ideas. No doubt earlier thinkers in other ancient societies played with them too. The ideas all bog down in the idea of God. Outside of God-answers there are few answers. Most of these are circular, as that there is something because we are here and a chain of cause and effect backs up to a first cause and that first cause could not come from nothing. That begs the question and makes us ask why there was a first cause rather than not, as well as why something can't come from nothing.

The new physics has put forth its own answer in the *anthropic principle*. It gives a me answer to the why question. Stephen Hawking states it this way: "We see the universe the way it is because if it were different, we would not be here to observe it." The universe is as it is so that we would be here. We might not be here if the universe were some other way. Nothing is some other way. Our asking, our being, makes the universe be here just as it is. It is so that I am.

This answer flirts with circularity. But it is fun to see physicists talk about it and ignore the philosophers who say they can't talk about it. The answer is weak since it depends on what may be a rare event in the universe: the existence of us with our ability to ask why questions.

Even if the carpet of the universe unrolled just for us, we have just come here in the last million years or so. The universe has grown out from the Big Bang for about 15 billion years. What if some dumb thing or creature asked, or some random group of atoms or stars spelled, the question "Why is there something?" *before* we arrived? Maybe the universe unfolds not for us but for smart things yet to evolve far across the cosmos.

The anthropic answer has a problem with the trillions or quadrillions or more branches of the universe's evolution tree. The universe can unfold in many ways. One branch or path took the world from the Big Bang to us. Many paths may lead to life forms like us. Far more paths may lead to smart life forms that we would not recognize if we saw or heard or felt them. They may be shouting at us right now. It may turn out that in time life sprouts in most universes that expand and that it sprouts as weeds sprout in rich damp soil. Life may mark an early growth phase of universes that grow. Most stars as they grow from young to old have a "water window" of a few billion years or so when orbiting planets could hold water on their surfaces. Be that as it may, the math says there could be or are or have been a vast number of other universes. In many of these a smart thing or being can ask the why-something question and assert the anthropic principle as an answer. That should make all those possible worlds exist just as it made our world exist. Here is one such world: the universe that equals ours but has one less human in it or even one less molecule in it. Why just one world? Why did we get lucky?

Maybe we didn't. Maybe our universe sits in a meta-universe along with zillions of others and all have, have had, or will have smart-life things in them. I think the anthropic principle leads to this. It acts like the tall black monolith in the movie *2001: A Space Odyssey*. It acts like a door that opens for the smart and stays closed for the dumb. Each world line or path has a smart thing or being in it or not. Those with a smart thing come to be. Those without never do. The anthropic principle filters or prunes the world lines and nips the dumb ones in the bud.

The anthropic principle opens a Pandora's box of smart worlds when it tries to explain just one. And it still does not explain why just asking a question or having a brain or brainlike thing picks out a world line. It gives no mechanism for how it picks or prunes a world line. We might as well say the universe is here because this rock or star is here and it would not be here if the universe did not unfold to make it so. The claim does not need life or smart things. It just needs things. The world unfolds for the thing. Why does it have to? For the same reason it has to unfold for us to be here. Why is that? Because we are here. And so on.

A FUZZY ANSWER: SUPPOSE NOTHING

I have a fuzzy answer to why there is something rather than nothing. The answer is this: If nothing, math blows up.

The answer may be deep or just math trivia. Either way it sets up a view of the world I want to explore. It deals with information. It deals with entropy, fuzzy entropy.

The answer comes from the math of fuzzy entropy. It comes when we ask a pure fuzzy question: How fuzzy is the universe? That answer falls between 0% and 100%. Fuzziness is a matter of degree. So the real question is whether the universe is a fuzzy set. If not, if the universe is just a set, the fuzz is 0%. Else it is more.

Most of us think the answer is 0%. We think the universe is not fuzzy. Each thing in the universe belongs here 100%. Things that don't belong in it belong 0%. The thing exists or not. Not in between. No shades of gray for existence.

The universe is the set of its objects. More than that: the universe is the set of all subsets of the universe. If it contains all things, it contains all sets of things too. And we think these pieces are black or white. A chunk of stuff belongs to the universe or not. Sets of stuff belong all or none.

Maybe. But logic can't tell us. It depends on test. The question is factual or empirical. Sets of things may well belong to the universe only to some degree. I think an "electron" or electron cloud belongs to a region of space only to some degree. But that is not the point. You don't have to know if the universe is fuzzy to answer why there is something rather than nothing. You just have to ask if it is fuzzy.

I first asked that one day in May 1991 as I shot archery in the large archery park near my home. The semester was over and I had stopped off on my way in to USC. I do that sometimes when the commute is bad. I was shooting six arrows at 40 yards at an elk target tied to three bales of straw. No one else was in the park. It was hot and dry and dusty and soon I forgot about the traffic jam and USC. Alone you can get "in" to archery in the Zen sense. That means you focus only on the task at hand and you focus so much that soon you do the task without focusing or trying. Then the mind frees up and relaxes. From time to time you catch yourself in this relaxed state and start to think about it and that can kill it. This time I caught myself thinking about why there is something rather than nothing. I don't know why it hit me then. I had just been hiking and trout fishing in the high Sierra Nevadas and started to think about it up there. Maybe my mind had since found an answer and this relaxed state let it bubble up. There it was right in front of my mind's eye. I could see the math symbols behind the idea.

The idea was this: Suppose there is nothing. Suppose there is not one thing. How do you do that? If you just work with words, you can get stuck as the old Greeks did with vague ideas of something from nothing and all that. I supposed nothing in a set sense. What set is nothing? We call it the *empty set* or the null set and write it as $\emptyset$. So what is nothing? What thing did I suppose was nothing? The world. The universe. The "space" of all things. We write that as X. To suppose nothing is to say that the space X is the empty set: $X = \emptyset$. That's the math talk. In that form I could see at once where it leads. It leads to a math explosion.

The fuzzy entropy theorem then says the entropy or fuzziness or vagueness of the universe equals zero over zero or 0/0.* The term 0/0 does not equal zero. And the two 0s don't cancel and give a 1. The term is undefined. We can't divide by zero or math blows up in a singularity. The assumption that leads to the blowup is that there is nothing. So we have to deny that. So there is something. That's my answer.

---

*The fuzzy entropy theorem says that the fuzziness of X is $E(X) = c(X \cap X^c)/c(X \cup X^c)$. Here c(A) just counts the elements or bits or fits in A. The opposite of everything is nothing. So $X^c = \emptyset$ always holds. So $X \cap X^c = \emptyset$ always holds. Now suppose there is nothing: $X = \emptyset$. Then $X \cup X^c = \emptyset$ holds too. So $E(X) = c(\emptyset)/c(\emptyset) = 0/0$. Impossible. So $X \neq \emptyset$. So something. Q.E.D.

You can see it in terms of fuzzy cubes. When there is nothing Aristotle collides with the Buddha. The cube corners implode to the midpoint. The midpoint explodes to the corners. Look at the case when there is just one thing left. Then the fuzzy cube is the line segment in Figure 15.1.

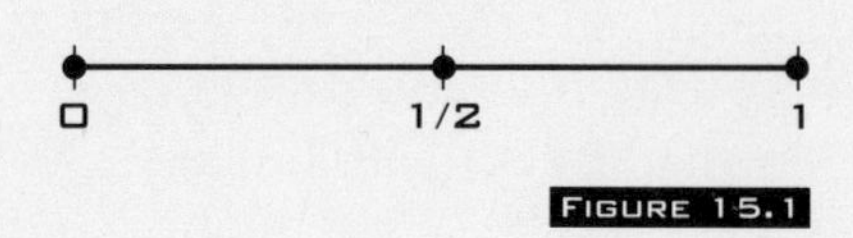

FIGURE 15.1

When you go from one last thing to no thing, from something to nothing, the line segment collapses into an infinitesimal point.

FIGURE 15.2

That is the fuzzy cube of zero dimensions. The Aristotle corners where A OR not-A holds 100% fall into the Buddha midpoint where A AND not-A holds 100%. The yin-yang equation ($A = A^c$) holds. Yet A AND not-A are distinct in the bivalent sense. Matter takes math with it down the drain.

That's the strange part. Going from one thing to no thing is a physical thing. The universe contains one last atom and then contains nothing. That is a matter of fact.

You may not like this answer. You may want to deny it on some technicality. But the math of it is simple. When there is nothing math blows up. I don't know how serious this is. But I like that it makes you think about how you might test the claim.

In theory we can test it. We just have to take all matter out of the universe. We don't know how to do that or where we would put the stuff or where we would put us to test the claim. But you can form the idea in your mind. A thought experiment. The last atom or photon or ball of stuff hangs in empty space and then disappears or implodes down to nothing. Maybe it disappears in a worm hole and leaves the old universe empty. You might have heard of that before. Gravity collapse.

The gravitational collapse of the universe could test the claim. It may test it. Say the universe has in it enough dark matter or neutrinos or any type of particle or wave that in time slows then halts the outward expansion from the Big Bang and then makes the universe fall back on itself and contract to the Big Crunch. That may well happen. A lot of us are pulling for it to happen. We don't want the universe to expand on and on and just peter out in a heat death. It takes about three electrons per cubic meter of space to make the universe collapse. It takes about 100 ghostly neutrinos that have no or minuscule mass and very very little charge.

No one knows what would happen if the universe fell back to the Big Crunch. The laws of physics break down at that singularity. So here we stand outside of science when we talk.

There is the view of the oscillating universe. The Big Crunch may be or may give rise to a new Big Bang and all grows again and repeats. Or the universe ball may crunch into its own black hole. It may get so small that it falls in a wormhole neck to some other universe where it pops out as a white hole or a big or small big bang or something stranger. In that case the old universe would go from something to nothing at least at its final instant. I would count that as a test. Physicists claim our current expansion will continue for about 10 billion years. A collapse to the Big Crunch would take 10 billion years on top of that. So the claim would take about 20 billion years to test.

We could also have a vacuum test. The vacuum is not "empty." It is active and pregnant with quantum possibilities. Maybe we could clear out part of it and fence it off to give a closed region of nothingness. Maybe we can fence off a black hole and feed it no matter for a few thousand years. Who knows?

The question is what it would be like if there was no math as we know it. You could not add or multiply. The number 2 might equal 3. The notion of number might disappear. Physicists say that the laws of physics need not hold in a big crunch or in a black hole. The fuzzy claim says that math need not hold there either. That is a strange claim. And it suggests something stranger.

It suggests that maybe logic does not differ from fact. Maybe they connect. That's the point I want to explore. Logic *and* fact. Math *and* things. Connected.

COSMIC CHIPS AND GOD

Let's back up. Why is there something rather than nothing? One answer is that if nothing then you have math trouble. One case of that is the case of fuzzy entropy, the mathematical measure of the fuzziness of an event or set. But so what? Why should the world care if some symbols foul up and no longer cohere?

The reason is the world seems to obey math. But it need not. It seems to but it does not have to. You can think of all our known math as a bush. It grows from a few roots and each day grows more branches. You can think of all science, all known fact, as a bush too. It too grows from a few key roots like gravity and light and molecular bonding and each day grows more branches, even new fact branches that don't fit yet to any other fact branch.

The two bushes are not the same. They grow with different laws. The math bush grows by pure deduction. The fact bush grows by tests and measurements in fits and starts—by induction. Scientists prune old branches and grow and graft new branches. New data guide the shears. But the two bushes are similar too. They have the same rough shape, at least in some clumps of branches. From a logical point of view there is no good reason for that. Logic moves in one direction. Facts can go in any direction. You never know in advance. That is why every test of fact is a gamble.

Science tracks math. Induction tracks deduction. As time passes the fact bush looks more and more like the math bush, branch for branch and twig for twig. The time lag makes clear which bush chases which.

Look at some of the great cases. In the last century James Clerk Maxwell found the four "Maxwell equations" of electricity and magnetism. Tests had confirmed them or pieces of them. Maxwell played with the math of the equations and out popped the wave theory of light. The math said light was a form of electromagnetism. The tests later confirmed that too. A few years later Einstein played with the math of relativity and out came the energy-mass equation $e = mc^2$. Later tests and nuclear bombs confirmed that too. It could have been otherwise. The tests could have found that $e = mc^5$ or $e = m^2c$ or an infinite number of other possible outcomes. But they didn't. The tests confirmed the math branch.

A few years later Einstein put forth his math of general relativity that said gravity is an illusion. Matter curves space. Energy and momentum curve the space-time continuum. The planet does not "attract" the meteoroid. The meteoroid flies by but rolls into the planet in some sense. It just looks like attraction. Einstein's curvature equations, like Maxwell's equations, lead to a wave equation. So the equations have solutions that radiate. That means that gravitational waves exist that travel at the speed of light. Over 70 years since then physicists have found indirect evidence for gravity waves in the orbital emissions of massive pulsars or neutron stars. In 1917, just after Einstein published his gravity or curvature equations, Karl Schwarzschild found a solution to them for a special symmetric case. His equations show that too much mass can make a gravity equation blow up to infinity (in effect a term gets divided by zero). This led to the prediction of black holes. Years later more and more evidence has suggested that black holes exist and tend to lie at the center of dense galaxies, including our Milky Way galaxy.

These are famous cases when fact followed math. To a lesser extent fact follows math every day in science in all fields. The more math we learn, the more of nature we see. The more nonlinear the math we learn, the more nonlinear nature seems to be. We ignored chaos as noise for hundreds of years. Recently we learned chaos math and recently we found chaos in weather patterns and heartbeats and molecular vibrations. The more we unravel the branches of the math bush the more relations we predict. In time the tests tend to confirm them. It could always go some other way but it does not. It hits and misses but overall it keeps going the way of math.

So where is God in all this? We see deeper and deeper into nature and we find no sign of Him. *No evidence*. No God in math. No God in fact. We have not seen or measured Him with microscope or telescope. He does not seem to be in the observable universe. And He seems to have left no footprints. We find only the smooth flow of events according to physical law. A flows into not-A and not-A flows into something else. What we can explain with God we can explain without Him.

Once again reason seems to end in doubt. There is the Hemingway challenge from the Fuzzy Past section. No one has produced a fact that is 100% true or 100% false. In the last chapter we saw the same sort of thing with ethics. No one has produced a pure moral claim that is true or false. Morals and the social contract do not seem to rest on logic or fact, fuzzy as those two may be. In the same way there seems no evidence for a God. The more we find out the more the ground seems to drop out from under us. And it is a short step from here to wonder why we fight to live on and propagate our genes and ideas. It all seems to lead to nihilism.

And it may end in nihilism. There may be no meaning or purpose to the world at least in a sense that we can grasp. Our talk of God may just be, as Pavlov said, a social reflex or, as Spinoza said, nature-inspired awe or, as Marx said, the opiate of the masses or, as Freud said, our own father deified into cosmic gas or, as the sociobiologists say, just some of our selfish genes that favor blind obedience to authority.

We feel we recognize God in the world or in ourselves but that may be an illusion. We recognize but cannot define. The neural nets in our brains are good at that. They evolved over hundreds of millions of years to do that, to quickly and ceaselessly match sensed patterns to stored patterns. We recognize faces and music and seasons and we have little or no idea how to define them. We cannot explain how we recall a name or answer a question or have a new idea. We just do it. Our neural nets just do it. And they may just be recognizing a God pattern where there is none. There seems to be no selective advantage in a genetic sense for glimpsing God. These God glimpses or the feeling of God recognition may be just a "filling in" or *déjà-vu* type anomaly of our neural nets. We may think we recognize God even though He does not exist just as we recognize the Kanizsa square in Figure 15.3 though it does not exist:

FIGURE 15.3

The neural nets in our eyes and brain make and sustain the Kanizsa-square illusion with its false boundaries and bright interior. It is not there on the page. It is not a Kantian noumenon or "thing in itself" out there beyond the senses. It is a phenomenon in our senses and brain. And our vague glimpses of God or His Shadow or His Handiwork may have the same status—an illusion in the neural wiring of a creature recently and narrowly evolved on a fluke of a planet in a fluke of a galaxy in a fluke of a universe.

I draw a different conclusion. *The universe is information.* Something like a big computer chip. I think someday we will find that energy connects with information. There may be information waves or particles or *infotons*. Information may be quantized in smart little infinitesimal particles like Leibniz's monads.

The more we look at nature the more information we see in the structure. The structure is the information. Our DNA is just genetic information made flesh. The neural nets in our brain and spine and muscles code and store and decode information. Our cultures and economies are just stores and flows of information. In the 1940s Claude Shannon at Bell Labs found the first "laws" of pure information theory. The world seems to obey them. The entropy of thermodynamics is the same as the entropy of abstract information theory. What we call a "pattern"—a face or a star or a galactic cluster—tends to be a local point of maximum information or minimum entropy. In 1957 physicist E. T. Jaynes at Stanford showed that the basic math law of statistical quantum mechanics (the Gibbs probability distribution) follows from maximizing the entropy of information theory. You need no data or tests or Niels Bohr to derive it. You just need abstract information theory.

What we seem to have just glimpsed is information. It started with bits. Now fuzzy logic has taken us further to fits. We have been working in large cubes and jumping from binary corner to binary corner. Now fuzzy math says there is a whole world inside the cube and inside the cube we are headed. We can dig through the cube from one black-and-white corner to another. Fuzzy math. Fuzzy physics. Fuzzy machine intelligence. We can replace probability and relative frequency with subsethood or the whole in the part. We can turn loose ever bigger neural nets and computers and linked neuro-computer networks to see more math and find more structure and gain more information. And this will be only the start: *It from fit*.

That raises the next question: Is God information? That may not be as strange as it sounds. We have made Him just about everything else: love, power, mind, energy, nature, maximum probability. But I don't think God as information is right or even makes sense. The universe is information. The physical structure is information. The universe is to God as the eye is to sight.

I think God has to do with how science tracks math when it does not have to. There I think we recognize Something we cannot define. We recognize a blueprint. Or Blueprint. With each new math insight and each new fuzzy fact we estimate a blueprint or math structure.

This may all change in the next second. The science bush may stop tracking the math bush. Fact may stop tracking logic. Energy may no longer equal mass times the square of the speed of light. Ocean waves may no longer move as the equations of fluid mechanics say they will move. Light bulbs may brighten or dwindle in intensity. The whole fabric of cause and effect may melt or unravel. Then this thesis melts too.

But suppose it does not. Suppose science keeps tracking math for hundreds or thousands or millions or billions of years. The blueprint hypothesis will grow in fuzzy truth. The Pythagorean theorem and all the other theorems will still give orders and we will still take them. The sense of shadow, the glimpses, the pattern we recognize but cannot define, will pass into clear view: There may be no God but the Mathmaker, and Science is His Prophet.

God is He who wrote the math. Or She who wrote the math. Or It that wrote the math. Or the Nothingness that wrote the math. The Mathmaker.

MAN A FUZZY MACHINE: THE RISE OF MACHINE IQs

What will be the quality of life in the fuzzy future? Man will live eons before he has something to say or not say to the Mathmaker. In the meanwhile higher machine IQs will change how we live and think and play. Fuzzy logic has given us our first small taste of smarter gadgets and the taste is sweet. Someday that may change.

What comes of man and his character as future high-MIQ systems drain his brain to theirs? What happens to hard work and dedication and moderation when high-MIQ systems make us so productive that we are all rich? Voice-in, voice-out computers wherever you turn. Fuzzy personality-profile chips in your smart answering machines. Smart prostheses wherever you need them. Platoons of smart cars on smart roads. Machine healthy and machine wealthy and machine wise.

What will it be like when everything you can do or think or create a smart machine can do better, far better? Will we live all year on vacation? Will we grow soft? Will each generation take fewer risks, meet fewer people, rely more on the state and large firms, spend more time in virtual-reality cybersuits and cyberchairs? Will man the rational animal end up an ultra-hightech couch potato?

I think things will turn out well because we will not get machine-rich all at once. Winning a big lottery can ruin you. Imagine how medieval kings and queens might have predicted the effects on us of cars, computers, airplanes, television, dentists, and personal freedom and enough economic means to use it. We will get pay raises and bonuses, not lottery jackpots, and we will know what to do with them. We will grow into our high-

MIQ toys and world gradually. As two cases in point consider the social impact of smart drugs and smart weapons.

Imagine what will happen when the first real high-IQ pills hit the streets. At first governments may outlaw them. They may outlaw them both for personal and professional use. Nobody can have high-IQ pills if everybody cannot have them. Leads to headaches or heart attacks or strokes or tumors in lab rats or computer-simulated rats. Long-term effects unknown. Prohibition will not stop the spread of high-IQ pills. It will only raise the price and that in turn will keep up the supply. It will encourage more chemists, graduate students, and teenagers to synthesize and sell the smart drugs on the black market. In time high-IQ pills will be legal. But legal or not they will change society. No one will want to be left out of the high-IQ race. High-IQ pills (or HIQs) are steroids for the mind. Everyone will take a HIQ. Students will take them to pass quizzes, final exams, and placement exams and to make up for nights when they played instead of studied. Workers will take them to meet deadlines, speak to groups, fill the suggestions box, get raises, and most of all to keep up with everyone else who takes them. Lovers will take HIQs to impress each other when they court. Artists will take them to create new works. Scientists will take them to come up with new ideas and to shoot down old ones. Soldiers will take them to improve the odds. Lawyers will take them to outwit opposing counsel. Physicians will take them to improve diagnoses and cut down on malpractice suits. In general the smart will take HIQs to stay smart. The less smart will take them to get smart. We will adapt to HIQs as we have adapted to cars and telephones and TV sets and frozen foods. A higher-IQ society will not lurch forward. It will move forward gradually.

Consider smart weapons. High-IQ weapons may just as well stabilize world relations as destabilize them. Our first experiments with smart weapons in the 1991 Iraq War were mixed-to-positive. The cruise missiles and smart rockets had very low machine IQs. Still overnight higher machine IQ became the new ante in arms races. When all sides have supersmart weapons, and have gotten them gradually in pace with their neighbors, they might tend to cancel one another out. Each country will have something like its own Star Wars shield. In some cases there will be hot tempers and pinpoint strikes instead of diplomacy. But when each country has mutually assured defense they might get thicker skins. The danger lies in the first few big MIQ advances in smart weapons. The twenty-first century will no doubt see plenty of that. Gradual MIQ rises in smart weapons might even make our world a better place. They might even render mass war obsolete.

In any event I think more machine wealth will have one supreme effect. *Machine wealth will help take duress out of our lives*. Duress. Diminished capacity. Forced decisions. We don't trust a confession that the police beats out of a suspect. Force or the threat of force increases duress. So do most environmental factors. Here machine wealth might do the most good. Of course poverty is duress. Few would work or steal if they did not have to.

But short lifespan is duress too, perhaps the ultimate duress. The ticking clock constrains everything we say and do. You would not act the same way if you knew you could live for a thousand or a million years. You might be slower to have children if you had them at all. You might be slower to spend than save. And you might be much slower to litter the planet and solar system. A longer life means a longer stake in society and in Earth and in the solar system. You might care if we mine or rape Mars or thaw it and "terraform" it into an Earthlike eden.

The sexual appetite puts us under duress. Every man knows he does not feel the same about the world just before sex and just after it. And women seem to feel a similar gap. So much of courtship and culture play upon slaking this thirst. A short lifespan is the real culprit. Future advances in robotics and materials and cosmetics will no doubt lead to sex substitutes, maybe even sex cyborgs modeled after the pop stars of the day. The demand for sex substitutes and sex boosters may be the only demand to exceed the demands for higher IQ and longer life. The AIDS epidemic will only increase the demand and supply of safe sex substitutes. A sexier world may well be a better world. It would sure be more fun.

But what happens to romantic love when every man and woman owns or leases a robotic harem? Free sex need not imply free love. But free sex may help promote it. What is left if we subtract lust and gene propagation from romance? What remains when we strip away the subtle layers of duress? Maybe nothing remains. Or maybe real love remains and can grow to new heights and in new forms that we cannot now imagine.

Machine wealth will drive a machine culture that we also cannot now imagine. It will weave new tapestries of art and grow new webs of science and open new doors of sensory perception and cognitive skill.

The combined result, and all the art and science and culture and history that precede it, will form the bottom layer on mankind's coral reef in the cosmos. Thousands, millions, maybe billions of years from now our biological and machine descendants will still add to the coral reef a skeleton and a culture and a science at a time. Our final descendants may stand on the coral reef and greet other civilizations in the universe who stand on their coral reefs. We may lead the pack or the pack may not let us in.

Or we may stand alone on a vain tower of machine Babel and search the universe for a Mathmaker who does not answer. Or if the Mathmaker answers us, we may not like the message. Maybe our universe is just a big chip in the void. It might store or process information for some other culture. The information may lie in how the universe changes. Patterns of cosmic expansion and contraction may encode or decode information as if our universe were a large neural net or compact disc or memory chip and we were only a viral colony squatting on one of the cold hard wires of one of the logic circuits. Or there may be message and nonmessage. The messages may be here now but we cannot perceive or conceive them. We may be like the ants that walk across the page of a calculus text.

Higher machine IQ opens the door to these worlds. Fuzzy logic has shown us that we crack open this door at a price. We have to disobey the old logic and break its laws to transcend it. When we first tried to get machines to think like us, we tried to think like them. We tried to think the way simple on-off machines worked. A cultural legacy of bivalence made this seem natural and proper. We looked for precision and provided it when we did not find it.

As we crack open the door further we may well have to abandon fuzzy logic in favor of some more general idea or theory or process. In the end fuzzy logic gives only a slightly better approximation of the truth as we have so far searched for it. Our science and math have just been born.

Man's future with smart machines will involve new variations on the old theme of master and slave. We will control our machine-intelligent superiors. We will live with them, create with them, adapt with them, maybe even reproduce with them. We will hold their strings while they hold ours. Both masters and both slaves. The question will be to what degree.

Glossary

adaptive fuzzy system. A fuzzy system that learns its rules from data. A human expert does not tell the system what the rules are. A stream of data feeds into a neural or statistical system and out come the fuzzy rules. An adaptive fuzzy system acts as a human expert. It learns from experience and uses fresh data to tune its stock of knowledge.

artificial intelligence (AI). The view of those in computer science that minds are computers. Popularized by HAL the computer in the movie *2001: A Space Odyssey.* AI views the mind as a type of symbol processor that works with strings of text or symbols much as a computer works with strings of 0s and 1s. In practice AI means expert systems or decision trees of bivalent rules. AI is the conceptual and political rival to the neural-network view of the brain. The AI crowd calls the neural view "connectionism."

associative memory. A system that stores data in parallel and searches for data or "recalls" data based on some feature of the data.

To store pictures of human faces an associative memory would superimpose each image on the same memory medium. A neural associative memory would superimpose the images on the same web of synapses or web of resistors in a neural chip. An optical associative memory would superimpose them on the same hologram or other photorefractive medium. In this sense an associative memory is a distributed memory unlike a RAM or random access memory that stores data at precise memory locations. If the associative memory lost a few of its synapses or resistors or chunks of its hologram, it would still retain the images in the surviving memory units. As it loses more units, the associative memory would still remember most of the images but the images would fade and the memory might confuse images when it recalls them.

If you stored 100 face images in an associative memory and then showed it the image of a new face, it would match the new face to the closest stored face. Some speech recognition systems store speech patterns in this way. The associative memory finds the match without an address look-up. From a dynamical systems point of view an associative memory stores patterns at the bottom of wells on a memory sheet or surface in the "space" of all patterns. Then a new pattern acts as a marble dropped on the memory surface and rolls down the well of the closest stored pattern.

bivalent logic. The logic most people mean when they say *logic*. Every statement or sentence is true or false or has the truth value 1 or 0. The simplest form is propositional logic, where each statement asserts a simple black-white description or "atomic fact" as in "The avocado is green." In predicate logic each statement asserts a set of black-white descriptions as in "All avocadoes are green." Aristotle first codified bivalent logic 300 years before the birth of Jesus. At the turn of the twentieth century Bertrand Russell and Alfred North Whitehead showed that most math reduces to bivalent logic.

chaos. An aperiodic equilibrium state of a dynamical system. A system in a chaotic equilibrium seems to wander "at random" through the states in some region of the state space. Yet the behavior is deterministic: A math equation describes it exactly. If you know the equation, you can predict in advance any point of the chaotic path or trajectory. Chaos has the property that if you pick any two starting points of a chaotic system, no matter how close, they give rise to two paths that will diverge in time. Most dynamical systems have chaotic equilibria. These range from the interaction of subatomic particles to the flutter of the olfactory bulb in your nose to the bubbles in a hot tub and the swirl of clouds in the sky to the distribution of galaxies in space.

cryonics. To freeze dead tissue with the hope of reviving it when it thaws. The early cryonics movement drew scorn from scientists and the press because it could not explain how future science might revive the unthawed dead. The freezing process damages tissue as the water between cells turns to ice and punctures the cells. In the 1980s the rise of nanotechnology showed how cryonics might work. Thaw a dead brain and then rebuild it a molecule at a time with tiny nano-robots or nanobots. As of 1993 there were over 30 patients in cryonic suspension. Most have suspended only their brains. The idea is that if nanotechnology can repair freezing damage and rejuvenate the dead brain, it can grow a lean young body from the head stump too.

Cryonics is legal in the United States but not in British Columbia, Canada. Organized resistance comes from morticians and county coroners and from cryobiologists who will not print cryonics articles in their "scientific" journals. Membership Article 2.04 of the Society of Cryobiology states that "the Board of Governors may refuse membership to applicants, or suspend or expel members . . . who promote any practice or application of freezing deceased persons in the anticipation of their reanimation."

digital signal processing. The branch of information science that converts analog measurements into lists of numbers or digits and then uses math to convert lists into new lists of numbers. Music on a compact disc comes from a sampled or digitized continuous signal.

DSP samples the signal 44,100 times per second and converts these samples to binary numbers. Signals can come from an earthquake or TV camera or deep-space probe. DSP can filter noise from a signal or enhance different parts of the signal. DSP commands on a compact-disc player can enhance different signal parts to give the effect of music played in a hall, church, jazz club, stadium, disco, or living room. In each case DSP converts the binary numbers on the optical disc to a new set of numbers that in turn make speakers vibrate in some special way.

dynamical system. A system that changes with time. In math a system described by a first-order differential or difference equation—a system whose velocity or rate of change is some function of time or of system parameters. In a broad sense everything is a dynamical system, the universe and all its pieces. The starting point of a dynamical system is an initial condition. The final point or points is the equilibrium state. In between lie the transient states. A leaf that falls from your hand to the ground is a dynamical system in a gravitational potential. The initial condition is where you let go of the leaf. The points the leaf falls through are the transient states. Where it lies on the ground is the steady state or fixed-point equilibrium. The leaf minimizes its potential energy as it falls. It starts out with maximum potential energy. As it falls the potential energy turns to kinetic energy and it follows a path of "least resistance" that ends in a potential energy minimum on the ground. All initial conditions lead to some minimum of potential energy on the ground.

A dynamical system can have two types of equilibrium states: periodic and aperiodic. Aperiodic equilibria are *chaotic* attractors. Once the system falls in one of these regions it moves around forever, or until something bumps it into a new state, with no apparent structure or periodicity to the movement. Many weather patterns are chaotic and so too it seems is the earth's slow movement or precession about its axis, which may in turn cause chaotic weather or climate.

The simplest periodic equilibrium is the *fixed point* attractor. The system stops at a fixed point. The state "ball" rolls down the attractor well and stops. Gravity brings objects to rest in fixed points of potential energy. Next up are *limit cycle* attractors where the state swirls round and round in equilibrium. Foxes and rabbits on an island end up in a limit cycle. As the rabbits grow in number so do the foxes who feed on them. The island can support only so many rabbits, and then rabbit population growth stops and crashes as the food runs out and as the many foxes catch and eat the rabbits. Then many of the foxes starve and that takes pressure off the few remaining rabbits, who have plenty to eat and whose population grows and then the cycle or some form of it repeats. Next up there are *limit torus* attractors. These equilibria act like a string that winds around a bagel in a drifting limit cycle. A limit cycle is a degenerate limit torus just as a fixed point is a degenerate limit cycle.

expert system. A search tree in artificial intelligence. An expert gives her knowledge as if-then rules and a programmer codes these in software. If the symptoms are so and so, the patient has typhus. If the patient has typhus and if this and that, then do such and such. Expert systems define a large logic tree or several small trees. The expert system has two pieces: the knowledge base and the inference engine. The knowledge base is just the tree or trees of bivalent rules.

The inference engine is some scheme for reasoning or "chaining" with the rules. In *forward chaining* you move from the root of the tree to a leaf of the tree an if-then step at a time. A forward chain is a prediction of effects given causes. It answers a what-if question. In *backward chaining* you move from a leaf or branch of a logic tree up to its root or as far back up as logic will take you. It searches for causes from known effects and so answers a why question.

Fuzzy systems are a type of expert system since they too store knowledge as rules—but as fuzzy rules or fuzzy patches. Expert systems work with black-white logic and symbols. Fuzzy systems work with fuzzy sets and have a numerical or math basis that permits both math analysis and simple chip design.

FAT theorem. Fuzzy approximation theorem. The FAT theorem shows that you can replace any system with a fuzzy system. In math terms the FAT theorem shows that a fuzzy system with a finite set of rules can uniformly approximate any continuous (or Borel-measurable) system. The system has a graph or curve in the space of all combinations of system inputs and outputs. Each fuzzy rule defines a patch in this space. The more uncertain the rule, the wider the patch. A finite number of small patches can always cover the curve. The fuzzy system averages patches that overlap. The key idea is that with enough napkins you can always cover any finite stretch of highway with the napkins.

fixed point attractor. A limit point of a dynamical system. The system falls into the attractor basin or "well" in the state space and stops. When the system enters the attractor region it acts as if it were a marble that rolls to the fixed point at the bottom of the attractor.

Neural nets encode patterns as the fixed points of neural dynamical systems. Then inputs or questions map to the nearest stored pattern. Some neural nets place computational solutions at fixed points. Then if you want to find a low-cost mix of resources or a low-cost schedule or route for several aircraft, you enter your resource data into the net and it swirls into the nearest low-cost solution. In theory you can always tie fixed points to the solutions of complex problems but in practice this is very hard to do. A fuzzy system can use a neural network to find fuzzy rules by tying data clusters to neural fixed points. The neural net might learn how a traffic cop associates traffic patterns with traffic commands. These associations appear as clusters in the data and as fixed points in the neural net.

fuzzy cognitive map (FCM). A fuzzy causal picture of the world. An FCM has concept nodes and causal edges. The concept nodes are fuzzy sets like "voter apathy" or "strength of the government." Each event belongs to or excites a concept node to some degree (most to zero degree). In the simplest case a concept node is just on or off and it acts as a threshold switch. If enough causal juice flows into it, it turns on. Else it turns off or stays off. In general a concept node "fires" or activates to some degree.

When a concept node fires it emits a type of causal juice that flows to the other FCM concept nodes across fuzzy causal edges. The edge or arrow is a fuzzy rule between fuzzy sets. The causal rule "Voter apathy increases the strength of the government" translates into an arrow from the voter-apathy node to the strength-of-government node. The edge is fuzzy because it can permit a small or large or other amount of causal juice to flow through it. The edge acts as a pipe of variable diameter through which the causal juice flows. An FCM edge can learn causal patterns by changing its effective pipe size as a function of how much causal juice flows through it or at what rate.

In practice an expert draws an FCM or a group of experts draws several FCMs. You can always combine any number of FCMs into one FCM. The final FCM graph of nodes and edges defines a nonlinear dynamical system that acts much as a neural network acts. You reason or predict with an FCM by turning on concept nodes like "voter apathy." Then the FCM net swirls and quickly settles down to some equilibrium state (fixed point, limit cycle, or chaotic attractor). The equilibrium state is the predicted outcome, the expected effect given the input cause. Engineers have applied FCMs to plant control, medical modeling, circuit analysis, and an array of social and political modeling.

fuzzy entropy. Measures the fuzziness of a fuzzy set or vague description. The measure ranges from 0% for a nonfuzzy or black-white set to 100% for a purely fuzzy set that equals its own opposite (midpoint of a fuzzy cube). The more a fuzzy set or description resembles its own opposite, the fuzzier it is and the greater is its fuzzy entropy.

In a fuzzy hypercube, where each point defines a fuzzy set, the fuzzy entropy of fuzzy set A is a simple ratio. It equals the distance of A from the nearest corner divided by the distance of A from the farthest corner. The Fuzzy Entropy Theorem says this ratio equals the count or size of the Buddha set A AND not-A divided by the count or size of the Aristotle set A OR not-A.

fuzzy logic. Has two meanings. The first meaning is multivalued or "vague" logic. Everything is a matter of degree including truth and set membership. This dates back to the turn of the century. The second meaning is reasoning with fuzzy sets or with sets of fuzzy rules. This dates back to the first work on fuzzy sets in the 1960s and 1970s by Lotfi Zadeh at the University of California at Berkeley. Zadeh chose the adjective "fuzzy" over the traditional adjective "vague" in his 1965 paper "Fuzzy Sets," and the name has stuck. Other synonyms: gray logic, cloudy logic, continuous logic.

fuzzy rule. A conditional of the form *IF X IS A, THEN Y IS B.* A and B are fuzzy sets: "IF the room air is COOL, THEN set the motor speed to SLOW." In math terms a rule is a relation between fuzzy sets. Each rule defines a fuzzy patch (the product A × B) in the system "state space"—the set of all possible combinations of inputs and outputs. The wider the fuzzy sets A and B, the wider and more uncertain the fuzzy patch. More certain knowledge leads to smaller patches or more precise rules. Fuzzy rules are the knowledge building blocks in a fuzzy system.

In math terms each fuzzy rule acts as an associative memory that associates the fuzzy response B with the fuzzy stimulus A. Then stimuli similar to A map to responses similar to B. In this sense each fuzzy rule defines fuzzy associative memory, or FAM. A set of FAM rules in a fuzzy system acts as a FAM at higher level. It too converts similar inputs to similar outputs.

fuzzy set. A set whose members belong to it to some degree. In contrast a standard or nonfuzzy set contains its members all or none. The set of even numbers has no fuzzy members. Each number belongs to it 0% or 100%. The set of big molecules has graded membership. Some molecules are bigger than others and so belong to it to greater degree. In the same way most properties like redness or tallness or goodness admit degress and thus define fuzzy sets.

In math terms a fuzzy set is either a point in a hypercube or a curve. A fuzzy set with n members is equal to a list of n numbers or *fit values*. Each fit value lies in the interval from 0 to 1 and stands for the degree that that member belongs to or fits in the fuzzy set. The set of all such lists of n fit values defines a solid unit hypercube of n dimensions (with

2^n corners made up of the 2^n binary lists of 0s and 1s or the 2^n nonfuzzy sets). Each fuzzy set is one point in this fuzzy cube. The same holds as the number n grows to infinity. In the three-dimensional case of three tall men the fit list (.9 .5 .3) means that the first man is 90% tall. The second man is 50% tall or as much not-tall as he is tall. The third man is 30% tall or more not-tall than tall.

A curve defines a fuzzy set for a continuum of cases like all possible temperature values between 50° and 100° or all possible car velocities between 0 mph and 120 mph. The height of the curve between 0 and 1 measures the fit value or degree that the element belongs to the fuzzy set. A nonfuzzy set looks like a step. Part of the curve is the flat line at 100% and the rest is the flat line at 0%. In this world continuity is a useful fiction for math analysis and for engineering design. Up close there are only discrete values and a finite and small set of temperature values or even car velocities. This amounts to "sampling" a fuzzy curve at several places and gives a finite fit list for the fuzzy set. The more samples the more accurate the fit list and the larger the dimension of the hypercube in which it sits as a point.

fuzzy system. A set of fuzzy rules that converts inputs to outputs. In the simplest case an expert states the rules in words or symbols. In the more complex case a neural system learns the rules from data or from watching the behavior of human experts. Each input to the fuzzy system fires all the rules to some degree as in a massive associative memory. The closer the input matches the if-part of a fuzzy rule, the more the then-part fires. The fuzzy system adds up all these output or then-part fuzzy sets and takes their average or centroid value. The centroid is the output of the fuzzy system. Fuzzy chips perform this associative mapping from input to output thousands or millions of times per second. Each map from input to output defines one FLIPS—or fuzzy logical inferences per second.

The Fuzzy Approximation Theorem (FAT) shows that a fuzzy system can model *any* continuous system. Each rule of the fuzzy system acts as a fuzzy patch that the system places so as to resemble the response of the continuous system to all possible inputs.

Kalman filter. An optimal math algorithm to predict the present state of a linear system given a fresh sensor measurement and all past sensor measurements. An optimal predictor gives an optimal estimate of the next or future state of the system given all present and past measurements of the system. A common application in aerospace is to estimate the position of an airplane or missile or satellite given imperfect measurements. Rudolph Kalman of the University of Florida at Gainesville first derived the Kalman filter in 1960 by a new application of the Pythagorean Theorem.

liar paradox. A paradox of self-reference. In ancient Greece the liar from Crete said that all Cretans are liars. Did he lie? If he did, then he told the truth and did not lie. If he did not lie, then he did. The two outcomes of a liar paradox have the contradictory form A AND not-A. This poses a problem or "paradox" for bivalent logic where either A is 100% true and not-A is 0% true or A is 0% true and not-A is 100% true. One bivalent contradiction is enough to end all bivalent logic and math. It poses no fuzzy problem because when the Cretan both lies and does not lie, he does each only 50%. Liar paradoxes reduce to literal half-truths in a fuzzy or multivalued logic.

linear system. A system whose whole equals the sum of its parts. All other systems are *nonlinear*. To study a linear system you cut it into small pieces and then study the small pieces and then patch them back to give the whole system. In math terms a linear system looks like a smooth sheet of typing paper. A nonlinear system looks like a crumpled sheet. Quantum mechanics is a linear system. Quantum matter waves add up to give linear effects. Linearity also accounts for the uncertainty relations between some quantum "conjugate" variables like position and momentum or energy and time.

LTI system. Linear time invariant system. A linear system whose structure does not change with time. LTI systems are the basis of modern signal processing. A sampler picks a large but finite number of values from an analog signal like a warbling voice or a squiggling seismograph. This converts the continuous signal into a finite list of digits or numbers. Under very general conditions an LTI system can reconstruct the *entire* analog signal from the samples if the samples are spaced closely enough. LTI systems can also filter noise in the signal or amplify key terms in the signal.

LTI systems, which include quantum mechanics, give rise to an "uncertainty principle" between the time and frequency components of a speech or image or other signal. This relation comes from the Pythagorean Theorem built into the linear structure of the system.

logical positivism. The view that only the statements of science or math are "meaningful." French philosopher Auguste Comte founded positivism in the early 1800s with his book *Social Physics*. In the 1920s philosophers and scientists at the University of Vienna, under the influence of Bertrand Russell's symbolic logic and the tests of Albert Einstein's general theory of relativity, founded logical positivism. They claimed that the meaning of a sentence is "in its method of verification." If you could not verify or test a statement of fact or prove or disprove a statement of logic or math, the statement had no meaning. This principle of verification, itself not verifiable, was to eliminate ethics, metaphysics, and most of philosophy and religion. Problems of mind and body and free will and the like were mere "pseudo-problems." Over the decades logical positivism has lost ground in philosophy but gained it in the sciences.

nanotechnology. The study of matter and computers at the level of one billionth of a meter. Dr. Erik Drexler gave the field its name. Nanotech promises to stack and unstack matter a molecule at a time. In contrast microtech works at the level of a millionth of a meter with large blobs of molecules even in the smallest microprocessor. Some nanotech visions include viral-type swarms of smart nano-robots or *nanobots* that can clean teeth or kill cancer cells or build complex machines.

neural network. A system of neurons and synapses that converts inputs to outputs. Also called a *neurocomputer*. The neural network is a nonlinear dynamical system. Its equilibrium states can recall or recognize a stored pattern or can solve a mathematical or computational problem.

You train a neural net by showing it examples. You do not write a computer program or give a set of math equations. You can teach a neural net to recognize your face in photos by showing it photos with your face in it and "rewarding" the net with a positive feedback signal and then showing it photos without your face and "punishing" the net with a negative feedback signal.

The structure of a neural net is a graph of nodes and edges. The *neurons* are the nodes. Neural signals pour into a neuron and then, if enough pour in, the neuron turns on or turns on to some degree. The *synapses* are the edges that connect the neurons. The synapses in a neural net, and in a brain, store memory and pattern information.

A neural net acts as an associative memory that stores associations between inputs and outputs, stimuli and responses. In theory neural networks, like the fuzzy systems in the FAT theorem, can model any system if they use enough neurons. In practice neural nets work well at learning to separate pattern classes—to tell a sea mine from a sea rock or to tell a safe pap smear from an unsafe one or to tell your signature on a check from someone else's or to read the zip code on an envelope.

Fuzzy systems use neural systems to learn fuzzy rules from examples or to tune the rules. The net learns the fuzzy rules by adapting its dynamical structure. The rules emerge as the equilibrium states of the neural dynamical system.

ontology. The philosophy of what exists and why. Same as metaphysics in formal philosophy. Has one big question: "What is there?" Has one big answer: "Everything." Different schools arise when they argue the details. Plato said what there is are ideas, the pure forms of thought like redness and not just red things, the "one above the many." Spinoza said everything is the pure substance God that has the two properties of extension and thought, which in turn lead to matter and mind.

Today the dominant view is physicalism—there is just the space-time continuum of physics. This leaves little room for the many math objects of science. Where do we put prime numbers or the functions of calculus? So Harvard's Willard Van Orman Quine and others have added math and logic to the space-time continuum: What there is is the space-time continuum and the minimum amount of math and logic that it takes to describe it accurately. Of course each day science tries to describe new things and accuracy is a matter of degree. So each day there is a new world and it holds to only some degree. (Quine's actual motto: "A theory is committed to those and only those entities which the bound variables of the theory must be capable of referring in order that the affirmations made in the theory be true.")

probability. The mathematical theory of chance. A probability is a number assigned to an event. The larger the number the more "likely" the event will occur. In probability theory all uncertainty comes from an undefined "randomness" or "chance."

In math terms all probability numbers must add up to one. All events are bivalent. Either an event happens or not, in which case its opposite happens. The probability that either the event A happens or its opposite not-A happens is 100%. Events in probability theory are just the black-white sets of set theory. In this sense probability theory rests on bivalent logic.

Pythagorean theorem. A theorem on right triangles: the square of the length of the hypotenuse equals the sum of the squares of the lengths of the other two sides ($c^2 = a^2 + b^2$). Pythagoras, a Greek philosopher and cult leader of the sixth century B.C., first proved the theorem and with it found the first irrational numbers.

The Pythagorean theorem remains the most important relation in mathematics. The two perpendicular sides of its right triangle lead to the *orthogonality conditions* of optimality in modern statistics and engineering, including the optimal Kalman filter for estimating the state of a linear system. In fuzzy theory the Pythagorean theorem leads to the subsethood theorem that measures how much one fuzzy set contains another.

set theory. The study of sets or classes of objects. The set is the basic unit in math just as the symbol is the basic unit in logic. Logic and set theory make up the "foundations" of math. In theory all the symbols of advanced calculus and nuclear physics are just shorthand for the longhand of sets and logic.

Classical set theory does not acknowledge the fuzzy or multivalued set whose members belong to the set to some degree. Classical set theory is bivalent. Every set contains all members all or none. Bertrand Russell showed that this leads to a bivalent liar paradox: Is the set of all sets that are not members of themselves a member of itself? For almost a century mathematicians and logicians have put forth new bivalent axioms to try to prevent paradoxes and just as quickly new paradoxes have emerged.

sorites paradox. A paradox that arises from a chain of bivalent if-then statements. In ancient Greece the sorites appeared as Zeno's paradox of the sand heap. Remove a grain of sand and the sand heap is still a heap. But in time if you remove enough sand grains the sand heap is no longer a heap. So which sand grain converts the heap to a non-heap, converts A to not-A? The sorites paradox does not arise in fuzzy logic because as the sand heap shrinks so too falls the degree of truth that the remaining sand is a heap.

In math terms the truth value of each bivalent statement is 1 and so is the truth of the whole chain of reasoning, since $1 \times 1 \times 1 \times \ldots \times 1$ still equals 1. For fuzzy truth values the reasoning chain falls in truth value for each step in the reasoning process, since then it equals the product of numbers less than 1.

subsethood. The degree to which one set contains another set. In classical set theory a set has subsets all or none. In fuzzy logic it is a matter of degree. That means the subsethood or containment value can take any value between 0% and 100%.

The measure of subsethood comes from the Subsethood Theorem. This in turn arises from the perpendicular or "orthogonal" relations of the Pythagorean Theorem. The subsethood theorem gives a new way to view the probability of an event. It equals *the whole in the part*. The probability of the part or event is the degree to which the whole or the "space" of all events is contained in the part. This relation cannot hold if subsethood is not fuzzy and can take on only the extreme black-white values 0% and 100%.

uncertainty principle. A trade-off between two bell curves. As one bell curve grows narrower and thus more certain, the other grows wider and thus less certain. In physics uncertainty principles hold between position and velocity (momentum) and between time and energy and between other variables. If you know how fast an electron speeds along a line, then you do not know with certainty where it is along the line, and vice versa.

Werner Heisenberg published the popular quantum version of the uncertainty principle in the late 1920s. At about the same time R.V.L. Hartley and H. Nyquist of Bell Labs published the same relations in an information or time-frequency setting. In fact every linear time-invariant system has an uncertainty principle and these systems extend well outside the bounds of physics.

vague logic. Same as fuzzy logic. Vague logic has statements that are true to some degree between 0 and 1. So Aristotle's ''law'' of excluded middle, A OR not-A, need not hold 100%. And the ''contradiction,'' A AND not-A, can hold to a degree greater than 0%. The term ''vague'' comes from Bertrand Russell and his work on multivalued logic in the early part of the twentieth century. Jan Lukasiewicz worked out the first formal vague or ''fuzzy'' or multivalued logics in the 1920s and 1930s. Philosopher Max Black extended vague logic to vague sets in 1937.

virtual reality (VR). Sometimes called *artificial* reality or cyberspace. A VR is a computer world that tricks the senses or mind. A virtual glove might give you the feel of holding your hand in water or mud or honey. A VR cybersuit might make you feel as if you swam through water or mud or honey. VR grew out of cockpit simulators used to train pilots and may shape the home and office multimedia systems of the future. The idea of advanced VR systems as future substitutes for sex and drugs and classroom training is the stock and trade of modern science fiction or ''cyberpunk'' writing.

Bibliography

Abraham, R. H., and C. D. Shaw. *Dynamics—The Geometry of Behavior. Part Three: Global Behavior*. Santa Cruz, CA: Aerial Press, 1984.

Anderson, B. D. O., and J. B. Moore. *Optimal Filtering*. Englewood Cliffs, NJ: Prentice Hall, 1979.

Anderson, J. A., and E. Rosenfeld. *Neurocomputing: Foundations of Research*. Cambridge, MA: M.I.T. Press, 1988.

Aquinas, T. *Summa Theologica & Summa Contra Gentiles*. Edited by A. C. Pegis. New York: Random House, 1948.

Aristotle. *Ethics*. New York: Penguin Classics, 1955.

———. *The Organon*.

———. *The Politics*. New York: Penguin Classics, 1962.

Ayer, A. J. *Language, Truth, and Logic*. Gollancz, 1936.

———. *The Problem of Knowledge*. New York: Penguin Books, 1956.

———. *Logical Positivism*. New York: The Free Press, 1959.

———. *Freedom and Morality and Other Essays*. Oxford: Clarendon Press, 1984.

Axelrod, R. *Structure of Decision: The Cognitive Maps of Political Elites*. Princeton: Princeton University Press, 1976.

———. *The Evolution of Cooperation*. New York: Basic Books, 1984.

Birkhoff, G., and J. von Neumann. "The Logic of Quantum Mechanics." *Annals of Mathematics* 37 (October 1936): 823–43.

Black, M. "Vagueness: An Exercise in Logical Analysis." *Philosophy of Science* 4 (1937): 427–55.

Born, M., and E. Wolf. *Principles of Optics*. 5th ed. New York: Pergamon, 1975.

Buchanan, J. M., and G. Tullock. *The Calculus of Consent: Logical Foundations of Constitutional Democracy*. Ann Arbor: University of Michigan Press, 1965.

Buddha. *The Dhammapada*. Translated by I. Babbit. New York: New Directions, 1936.

———. *Teachings of the Compassionate Buddha*. Edited by E. A. Burtt (ed.), Penguin, New York, 1955.

———. *A Buddhist Bible*. Edited by D. Goddard. Boston: Beacon Press, 1966.

Carnap, R. "Testability and Meaning." *Philosophy of Science* 3 (1936).

———. *Meaning and Necessity*. Chicago: University of Chicago Press, 1947.

———. *Logical Foundations of Probability*. Chicago: University of Chicago Press, 1950.

———. *The Logical Structure of the World & Pseudoproblems in Philosophy*. Berkeley: University of California Press, 1967.

———. *Two Essays on Entropy*. Berkeley: University of California Press, 1977.

Carnap, R., and M. Gardner. *An Introduction to the Philosophy of Science.* New York: Basic Books, 1966.

Cheeseman, P. "In Defense of Probability." *Proceedings of the International Joint Conference on Artificial Intelligence.* (August 1985): 1002–1009.

Choquet, G. "Theory of Capacities." *Ann. Institut Fourier* 5 (1953): 131–295.

Chung, K. L. *A Course in Probability Theory.* Orlando, FL: Academic Press, 1974.

Churchland, P. M. *A Neurocomputational Perspective: The Nature of Mind and the Structure of Science.* Cambridge, MA: M.I.T. Press, 1989.

Cohen-Tannoudji, C., B. Diu, and F. Laloe. *Quantum Mechanics.* Vol. I. New York: Wiley-Interscience, 1977.

Comte, A. *Early Essays on Social Philosophy.* London: New Universal Library, 1911.

Copleston, F. C. *Aquinas.* New York: Penguin Books, 1955.

Cox, R. T. "Probability, Frequency, and Reasonable Expectations." *American Journal of Physics* 14 (January 1946): 1–13.

Crick, F. "Function of the Thalamic Reticular Complex: The Searchlight Hypothesis." *Proceedings of the National Academy of Sciences* 81 (1984): 4586–90.

Daugman, J. G. "Uncertainty Relation for Resolution in Space, Spatial Frequency, and Orientation Optimized by Two-Dimensional Visual Cortical Filters." *Journal of the Optical Society of America* 2 (July 1985).

Davidson, D. "The Structure and Content of Truth." *Journal of Philosophy* 87 (June 1990): 279–328.

Descartes, R. *Meditations.* 1641.

Dixon, R. C. *Spread Spectrum Systems.* 2d ed. New York: Wiley-Interscience, 1984.

Dowling, J. E. *Neurons and Networks: An Introduction to Neuroscience.* Cambridge, MA: Harvard University Press, 1992.

Drexler, K. E. *Engines of Creation: The Coming Era of Nanotechnology.* New York: Anchor Press, 1986.

Dunham, W. *Journey Through Genius: The Great Theorems of Mathematics.* New York: Wiley, 1990.

Dworkin, R. M. "Is Law a System of Rules?" In *Essays in Legal Philosophy,* edited by R. S. Summers. New York: Oxford University Press, 1968.

———. *Taking Rights Seriously.* Cambridge, MA: Harvard University Press, 1977.

Eddington, A. *The Philosophy of Physical Science.* Ann Arbor: University of Michigan Press, 1958.

Einstein, A. *The Principle of Relativity: A Collection of Original Papers on the Special and General Theory of Relativity.* New York: Dover, 1952.

Epicurus. *Letters, Principal Doctrines, and Vatican Sayings.* Indianapolis: Bobbs-Merrill, 1964.

Feyerabend, P. *Against Method.* London: New Left Books, 1975.

Fukunaga, K. *Statistical Pattern Recognition.* 2d ed. New York: Academic Press, 1990.

Gabor, D. "Theory of Communication." *Journal of the Institution of Electrical Engineers* 93 (1946): 429–57.

Gaines, B. R. "Foundations of Fuzzy Reasoning." *International Journal of Man-Machine Studies* 8 (1976): 623–88.

Gardner, M. *Science: Good, Bad and Bogus*. Oxford: Oxford University Press, 1982.

Gettier, E. "Is Justified True Belief Knowledge?" *Analysis* 23 (June 1963): 121–23.

Gibbins, P. *Particles and Paradoxes: The Limits of Quantum Logic*. Cambridge: Cambridge University Press, 1987.

Gibbons, G. W., S. W. Hawking, and S. T. C. Siklos. *The Very Early Universe*. Cambridge: Cambridge University Press, 1983.

Gibson, W. *Neuromancer*. New York: Ace Books, 1984.

Gilmore, R. *Catastrophe Theory for Scientists and Engineers*. New York: Wiley, 1981.

Goleman, D. *The Meditative Mind: The Varieties of Meditative Experience*. New York: St. Martin's Press, 1988.

Gotoh, K., J. Murakami, T. Yamaguchi, and Y. Yamanaka. "Application of Fuzzy Cognitive Maps to Support Plant Control" (in Japanese). *SICE Joint Symposium of the 15th Syst. Symp. and 10th Knowledge Engineering Symposium*. 1989: 99–104.

Gould, S. J. "Is a New and General Theory of Evolution Emerging?" *Paleobiology* 6, no. 1 (1980): 119–30.

Gray, R. M., and L. D. Davisson. *Random Processes: A Mathematical Approach for Engineers*. Englewood Cliffs, NJ: Prentice Hall, 1986.

Grice, H. P. "Meaning." *Philosophical Review* 66 (1957): 377–88.

Grossberg, S. *Studies of Mind and Brain*. Boston: Reidel, 1982.

Grover, D. "Truth and Language-World Connections." *Journal of Philosophy* 87 (December 1990).

Gudder, S. P. *Quantum Probability*. New York: Academic Press, 1988.

Gutman, R. W. *Richard Wagner: The Man, His Mind, and His Music*. New York: Time Books, 1968.

Hamming, R. W. *Digital Filters*. 2d ed. Englewood Cliffs, NJ: Prentice Hall, 1983.

———. *Coding and Information Theory*. 2d ed. Englewood Cliffs, NJ: Prentice Hall, 1986.

Hart, H. L. A. *The Concept of Law*. New York: Oxford University Press, 1961.

Hartley, R. V. L. "Transmission of Information." *The Bell Systems Technical Journal* 7 (1928): 535–63.

Hawking, S. W. *A Brief History of Time: From the Big Bang to Black Holes*. New York: Bantam Books, 1988.

Hawking, S. W., and G. F. R. Ellis. *The Large-Scale Structure of Space-Time*. Cambridge: Cambridge University Press, 1973.

Hayek, F. A. *The Counter-Revolution of Science: Studies on the Abuse of Reason*. Glencoe, IL: Free Press, 1952.

Hebb, D. O. *The Organization of Behavior*. New York: Wiley, 1949.

Heidegger, M. *An Introduction to Metaphysics*. New Haven: Yale University Press, 1959.

Heisenberg, W. *The Physical Principles of the Quantum Theory*. Chicago: University of Chicago Press, 1930.

———. *Physics and Philosophy: The Revolution in Modern Science*. New York: Harper & Row, 1958.

Hemingway, E. *A Moveable Feast*. New York: Collier Books, 1964.

Hempel, C. G. *Scientific Explanation: Essays in the Philosophy of Science*. New York: The Free Press, 1965.

Henderson, J. M., and R. E. Quandt. *Microeconomic Theory: A Mathematical Approach*. 3d ed. New York: McGraw-Hill, 1980.

Herrigel, E. *Zen in the Art of Archery*. New York: Vintage, 1953.

Hobbes, T. *Leviathan*. New York: Pocket Books, 1964.

Hopfield, J. J. "Neural Networks and Physical Systems with Emergent Collective Computational Abilities." *Proceedings of the National Academy of Sciences* 79 (1982): 2554–58.

Hopfield, J. J., J. N. Onuchic, and D. N. Beratan. "A Molecular Shift Register Based on Electron Transfer." *Science* 241 (12 August 1988): 817–20.

Hornik, K., M. Stinchcombe, and H. White. "Multilayer Feedforward Networks are Universal Approximators." *Neural Networks* 2 (1989): 359–66.

Huang, K. *Quarks, Leptons & Gauge Fields*. Singapore: World Scientific, 1982.

Hume, D. *An Inquiry Concerning Human Understanding*. 1748.

Hunter, G. *Metalogic*. Berkeley: University of California Press, 1973.

Jain, A. K. "Image Data Compression: A Review." *Proceedings of the IEEE* 69 (March 1981): 349–89.

Jaynes, E. T. "Where Do We Stand on Maximum Entropy?" In *The Maximum Entropy Formalism*, edited by Levine and Tribus. Cambridge, MA: M.I.T. Press, 1979.

Johnson, P. *Modern Times: The World From the Twenties to the Eighties*. New York: Harper & Row, 1983.

Jung, C. G. *Man and His Symbols*. London: Aldus Books, 1964.

Kac, M. *Probability and Related Topics in Physical Sciences: Lectures in Applied Mathematics*. Vol. 1. New York: Interscience, 1959.

———. *Statistical Independence in Probability Analysis and Number Theory*. The Mathematical Society of America, 1959.

Kailath, T. *Linear Systems*. Englewood Cliffs, NJ: Prentice Hall, 1980.

Kalman, R. E. "A New Approach to Linear Filtering and Prediction Problems." *Transactions ASME Journal Basic Engineering Series D* 82 (1960): 35–46.

Kanizsa, G. "Subjective Contours." *Scientific American* 234 (1976): 48–52.

Kant, I. *Prolegomena to Any Future Metaphysics*. New York: Liberal Arts Press, 1950.

———. *The Metaphysical Principles of Virtue*. New York: Bobbs-Merrill, 1964.

———. *Critique of Pure Reason*. 2d ed. 1787.

Kelsen, H. *General Theory of Law and State*. Cambridge, MA: Harvard University Press, 1954.

Keynes, J. M. *A Treatise on Probability*. London: Macmillan, 1921.

Kierkegaard, S. *Either-Or*. 1843.

Kinser, J. M., H. J. Caulfield, and J. Shamir. "Design for a Massive All-Optical Bidirectional Associative Memory: The Big BAM." *Applied Optics* 27, no. 16 (15 August 1988): 3442–43.

Kline, M. *Mathematics: The Loss of Certainty*. New York: Oxford University Press, 1980.

Kong, S., and B. Kosko. "Adaptive Fuzzy Systems for Backing up a Truck-and-Trailer," *IEEE Transactions on Neural Networks* 3, no. 2 (March 1992).

Kosko, B. "Fuzzy Cognitive Maps." *International Journal of Man-Machine Studies* 24 (January 1986): 65–75.

———. "Differential Hebbian Learning." *Proceedings American Institute of Physics: Neural Networks for Computing* (April 1986): 277–82.

———. "Fuzzy Entropy and Conditioning." *Information Sciences* 40 (1986): 165–74.

———. "Counting with Fuzzy Sets." *IEEE Transactions on Pattern Analysis and Machine Intelligence* 8 (July 1986): 556–57.

———. "Fuzzy Knowledge Combination." *International Journal of Intelligent Systems* 1, no. 4 (Winter 1986): 293–320.

———. "Adaptive Bidirectional Associative Memories." *Applied Optics* 26, no. 23 (1987): 4947–60.

———. *Foundations of Fuzzy Estimation Theory*. Ann Arbor: UMI, 1987.

———. "Bidirectional Associative Memories." *IEEE Transactions on Systems, Man, and Cybernetics* 18 (1988): 49–60.

———. "Hidden Patterns in Combined and Adaptive Knowledge Networks." *International Journal of Approximate Reasoning* 2, no. 4 (October 1988): 377–93.

———. "In Defense of God." *IEEE Expert* 5, no. 1 (February 1990): 74–76.

———. "Unsupervised Learning in Noise." *IEEE Transactions on Neural Networks* 1, no. 1 (March 1990): 44–57.

———. "Fuzziness vs. Probability." *International Journal of General Systems* 17, no. 2 (1990): 211–40.

———. "Equilibrium in Local Marijuana Games." *Journal of Social and Biological Structures* 14, no. 1 (1991): 51–65.

———. "Stochastic Competitive Learning." *IEEE Transactions on Neural Networks* 2, no. 5 (September 1991): 522–29.

———. "Smarter Weapons, Harder Fights." *Liberty* 5, no. 1 (September 1991): 37–38.

———. *Neural Networks and Fuzzy Systems: A Dynamical Systems Approach to Machine Intelligence*. Englewood Cliffs, NJ: Prentice Hall, 1992.

———. *Neural Networks for Signal Processing*. Englewood Cliffs, NJ: Prentice Hall, 1992.

———. "Fuzzy Systems as Universal Approximators." *IEEE Transactions on Computers*, 1993; *Proceedings of the 1992 IEEE Conference on Fuzzy Systems* (March 1992): 1153–62.

———. "Invest in Higher Machine IQ." *Reason* (October 1992): 42–44.

———. "Addition as Fuzzy Mutual Entropy." *Information Sciences* 73 (October 1993): 273–84.

———. "The Probability Monopoly." *International Journal of Approximate Reasoning*, 1993.

Kosko, B., and J. S. Limm. "Vision as Causal Activation and Association."

Proceedings SPIE: Intelligent Robots and Computer Vision 579 (September 1985): 104–109.

Kreuger, M. *Artificial Reality II*. 2d ed. Reading, MA: Addison-Wesley, 1991.

Kuhn, T. *The Structure of Scientific Revolutions*. Chicago: University of Chicago Press, 1962.

Kulikowski, J. J., S. Marcelja, and O. P. Bishop. "Theory of Spatial Position and Spatial Frequency Relations in the Receptive Fields of Simple Cells in the Visual Cortex." *Biological Cybernetics* 43 (1982): 187–98.

Lakoff, G. "Hedges: A Study in Meaning Criteria and the Logic of Fuzzy Concepts." *Journal of Philosophical Logic* 2 (1973): 458–508.

Lao-tze. *Tao Te Ching*. Edited by Feng and English. New York: Vantage Press, 1972.

Leggett, A. J. "Quantum Mechanics at the Macroscopic Level," In *Chance and Matter*, edited by F. Souletie, J. Vannimenus, and R. Stora, 395–506. Elsevier Science Publishers, 1987.

Leibniz, G. W. F. *Selected Writings*. Edited by P. P. Wiener. New York: Scribner's, 1951.

Lerner, R. G., and G. L. Trigg. *Encyclopedia of Physics*. 2d ed. VCH Publishers, 1990.

Lindley, D. V. "The Probability Approach to the Treatment of Uncertainty in Artificial Intelligence and Expert Systems." *Statistical Science* 2, no. 1 (February 1987): 17–24.

McKeon, R. *An Introduction to Aristotle: Selected Readings*. New York: Random House, 1947.

Mamdani, E. H. "Application of Fuzzy Logic to Approximate Reasoning Using Linguistic Synthesis." *IEEE Transactions on Computers* C-26 (December 1977): 1182–91.

Marr, D. *Vision*. San Francisco: W. H. Freeman, 1985.

Maynard-Smith, J. "The Theory of Games and the Evolution of Animal Conflicts." *Journal of Theoretical Biology* 47 (1974): 209–21.

———. *Evolution Now: A Century After Darwin*. San Francisco: W. H. Freeman, 1982.

———. *Evolution and the Theory of Games*. Cambridge: Cambridge University Press, 1982.

Mead, C. *Analog VLSI and Neural Systems*. Reading, MA: Addison-Wesley, 1989.

Mendelson, E. *Introduction to Mathematical Logic*. Princeton: Van Nostrand, 1964.

Mill, J. S. *A System of Logic*. 1843.

———. *On Liberty*. London: Longmans, Green, Reader, and Dyer, 1870.

Minsky, M. L. *Robotics*. New York: Anchor Press, 1985.

———. *The Society of Mind*. New York: Simon & Schuster, 1985.

Minsky, M. L., and S. Papert. *Perceptrons: An Introduction to Computational Geometry*. 2d ed. Cambridge, MA: M.I.T. Press, 1988.

von Mises, R. *Probability, Statistics, and Truth*. London: William Hodge, 1939.

———. *Positivism: A Study in Human Understanding*. Cambridge, MA: Harvard University Press, 1951.

Moore, A. W. *The Infinite*. New York: Routledge, 1990.

———. *Philosophical Studies*. London: Routledge, 1922.

Murphy, M., and S. Donovan. *The Physical and Psychological Effects of Meditation*. San Rafael, CA: Esalen Institute, 1988.

Musashi, M. *A Book of Five Rings*. Woodstock, NY: Overlook Press, 1974.

Nadelmann, E. A. "Drug Prohibition in the United States: Costs, Consequences, and Alternatives." *Science* 245 (1989): 939–47.

Nagel, Ernest. *The Structure of Science: Problems in the Logic of Scientific Explanation*. Hackett, 1979.

Needy, C. W., ed. *Classics of Economics*. Oak Park, IL: Moore Publishing, 1980.

Nilsson, N. *Learning Machines*. New York: McGraw-Hill, 1965.

Nyquist, H. "Certain Topics in Telegraph Transmission Theory." *AIEE Transactions* (1928): 617–44.

Notovich, N. *The Unknown Life of Jesus Christ*. Chicago: Rand McNally, 1894.

Oppenheim, A. V., and R. W. Schafer. *Discrete-Time Signal Processing*. Englewood Cliffs, NJ: Prentice Hall, 1989.

Oppenheimer, F. *The State*. New York: Viking Press, 1942.

Owen, G. *Game Theory*. 2d ed. New York: Academic Press, 1982.

Pacini, P. J., and B. Kosko. "Adaptive Fuzzy Systems for Target Tracking." *Intelligent Systems Engineering* 1, no. 1 (1992).

Papineau, D. *Theory and Meaning*. Oxford University Press, 1979.

Parker, T. S., and L. O. Chua. "Chaos: A Tutorial for Engineers." *Proceedings of the IEEE* 75 (August 1987): 982–1008.

Pauli, W. *Theory of Relativity*. New York: Pergamon Press, 1958.

Pauling, L. *General Chemistry*. San Francisco: W. H. Freeman and Company, 1970.

Pavlov, I. P. *Conditioned Reflexes*. New York: Oxford University Press, 1927.

Peirce, C. S. *Collected Papers*. Cambridge, MA: Harvard University Press, 1932.

Pelikan, J., ed. *The World Treasure of Modern Religious Thought*. Boston: Little and Brown, 1990.

Penrose, R. *The Emperor's New Mind: Concerning Computers, Minds, and the Laws of Physics*. New York: Penguin Books, 1989.

Plato. *The Republic*. New York: Penguin Classics, 1967.

Popper, K.R. *The Logic of Scientific Discovery*. New York: Basic Books, 1965.

Pratt, W. K. *Digital Image Processing*. New York: Wiley, 1978.

Prestowitz, C. V. *Trading Places: How We Are Giving Our Future to Japan and How to Reclaim It*. New York: Basic Books, 1989.

Prigogine, I. *From Being to Becoming*. San Francisco: W. H. Freeman, 1980.

Prophet, E. C. *The Lost Years of Jesus*. Livingston, MT: Summit University Press, 1984.

Quine, W. V. O. *Mathematical Logic*. Cambridge, MA: Harvard University Press, 1940.

———. "Two Dogmas of Empiricism." *Philosophical Review* 60 (1950): 20–43.

———. *Word and Object*. Cambridge, MA: M.I.T. Press, 1960.

———. *From a Logical Point of View: Nine Logico-Philosophical Essays*. 2d ed. New York: Harper Torchbooks, 1961.
———. *Set Theory and Its Logic*. Cambridge, MA: Harvard University Press, 1963.
———. *Ontological Relativity and Other Essays*. New York: Columbia University Press, 1969.
———. *Philosophy of Logic*. Englewood Cliffs, NJ: Prentice Hall, 1970.
———. *The Ways of Paradox and Other Essays*. 2d ed. Cambridge, MA: Harvard University Press, 1976.
———. *Theories and Things*. Cambridge, MA: Harvard University Press, 1981.
———. "What Price Bivalence?" *Journal of Philosophy* 78 (February 1981): 90–95.
———. "States of Mind." *Journal of Philosophy* 82 (January 1985): 5–8.
———. *Quiddities: An Intermittently Philosophical Dictionary*. Cambridge, MA: Harvard University Press, 1987.
Quine, W. V. O., and J. S. Ullian. *The Web of Belief*. 2d ed. New York: Random House, 1978.
Rawls, J. "Justice as Fairness," *Philosophical Review* (1955): 164–194.
———. *A Theory of Justice*. Cambridge, MA: Harvard University Press, 1971.
Reichenbach, H. *The Theory of Probability*. Berkeley: University of California Press, 1949.
———. *The Philosophy of Space and Time*. New York: Dover, 1957.
Resher, N. *Many-Valued Logic*. New York: McGraw-Hill, 1969.
Rosser, J. B., and A. R. Turquette. *Many-Valued Logics*. New York: North-Holland, 1952.
Rousseau, J. J. *The Social Contract*. 1762.
Rudin, W. *Real and Compless Analysis*. 2d ed. New York: McGraw-Hill, 1974.
Rumelhart, D. E., and J. L. McClelland. *Parallel Distributed Processing*. Vol 1. Cambridge, MA: M.I.T. Press, 1986.
Russell, B. *Introduction to Mathematical Philosophy*. New York: Simon & Schuster, 1917.
———. "Vagueness," *Australian Journal of Philosophy* 1 (1923).
———. *Mysticism and Logic*. New York: Doubleday/Anchor, 1927.
———. *A History of Western Philosophy*. New York: Simon & Schuster, 1945.
———. *Human Knowledge: Its Scope and Limits*. New York: Simon & Schuster, 1948.
———. *The Problems of Philosophy*. Oxford: Oxford University Press, 1959.
———. *The Basic Writings of Bertrand Russell*. Edited by Enger and Denonn. New York: Simon & Schuster, 1961.
———. *Philosophical Essays*. New York: Simon & Schuster, 1966.
———. *The Philosophy of Logical Atomism*. LaSalle, IL: Open Court, 1985.
Schrodinger, E. *Space-Time Structure*. Cambridge: Cambridge University Press, 1950.
Sellars, W., and J. Hospers. *Readings in Ethical Theory*. 2d ed. Englewood Cliffs, NJ: Prentice Hall, 1970.
Shafer, G. *A Mathematical Theory of Evidence*. Princeton: Princeton University Press, 1976.

Shannon, C. E. "The Mathematical Theory of Communication." *The Bell System Technical Journal* 27 (1948): 379–423.

Shannon, C. E., and W. Weaver. *The Mathematical Theory of Communication*. Urbana, IL: University of Illinois Press, 1949.

Shepherd, G. M. *The Synaptic Organization of the Brain*. 2d ed. New York: Springer-Verlag, 1979.

Sidgwick, H. *The Methods of Ethics*. Chicago: University of Chicago Press, 1962.

Skinner, B. F. *Science and Human Behavior*. New York: The Free Press, 1953.

Skyrms, B. "The Explication of 'X Knows That p.' " *Journal of Philosophy* 64 (June 1967): 373–89.

Smith, H. *The Religions of Man*. New York: Harper & Row, 1958.

Smuts, J. C. *Holism and Evolution*. London: Macmillan, 1926.

Snow, C. P. *The Two Cultures and the Scientific Revolution*. New York: Cambridge University Press, 1959.

Spooner, L. *No Treason: The Constitution of No Authority*. Colorado Springs, CO: Ralph Myles, 1973.

Stark, H., and J. W. Woods. *Probability, Random Processes, and Estimation Theory for Engineers*. Englewood Cliffs, NJ: Prentice Hall, 1986.

Steinbeck, J. *The Log from the* Sea of Cortez. New York: Penguin Books, 1951.

Strang, G. *Linear Algebra and Its Applications*. 3d ed. San Diego, CA: Harcourt Brace Jovanovich, 1988.

Sugeno, M. "An Introductory Survey of Fuzzy Control." *Information Sciences* 36 (1985): 59–83.

Suzuki, D. T. *An Introduction to Zen Buddhism*. New York: Grove Press, 1964.

Suzuki, S. *Zen Mind, Beginner's Mind*. New York: Weatherhill, 1970.

Taylor, A. E. *Aristotle*. New York: Dover, 1955.

Taber, W. R. "Knowledge Processing with Fuzzy Cognitive Maps." *Expert Systems with Applications* 2, no. 1 (1991): 83–87.

Tarski, A. *Logic, Semantics, Metamathematics*. Oxford: Clarendon, 1956.

Thom, R. *Structural Stability and Morphogenesis*. Reading, MA: Addison-Wesley, 1975.

Thompson, R. F. *The Brain: An Introduction to Neuroscience*. New York: W. H. Freeman, 1985.

———. "The Neurobiology of Learning and Memory." *Science* (29 August 1986): 941–47.

Togai, M., and H. Watanabe. "Expert System on a Chip: An Engine for Realtime Approximate Reasoning." *IEEE Expert* 1, no. 3 (1986).

Tsypkin, Y. Z. *Foundations of the Theory of Learning Systems*. Orlando, FL: Academic Press, 1973.

Upanishads: Breath of the Eternal. New York: Mentor, 1948.

Van Creved, M. *Technology and War*. Free Press, 1989.

Van Fraassen, B. C. *The Scientific Image*. Oxford: Oxford University Press, 1980.

———. *Quantum Mechanics: An Empiricist View*. Oxford: Clarendon Press, 1991.

Van Wolferen, K. *The Enigma of Japanese Power*. New York: Knopf, 1989.

Wald, R. M. *General Relativity*. Chicago: University of Chicago Press, 1984.

Watts, A. W. *The Way of Zen*. New York: Vintage Books, 1957.

Whitehead, A. N., and B. Russell. *Principia Mathematica*, 2d ed. Cambridge: Cambridge University Press, 1927.

Widrow, B., and S. D. Stearns. *Adaptive Signal Processing*. Englewood Cliffs, NJ: Prentice Hall, 1985.

Wiener, N. *Cybernetics: Control and Communication in the Animal and the Machine*. Cambridge, MA: M.I.T. Press, 1948.

Williams, W. E. "South Africa Is Changing." *San Diego Union*. Heritage Foundation Syndication, August 1986.

Wilson, E. O. *Sociobiology: The New Synthesis*. Cambridge, MA: Harvard University Press, 1975.

Winston, P. H. *Artificial Intelligence*. 2d ed. Reading, MA: Addison-Wesley, 1984.

Wittgenstein, L. *Tractatus Logico-Philosophicus*. London: Routledge & Kegan Paul Ltd., 1922.

Yamakawa, T. "A Simple Fuzzy Computer Hardware System Employing MIN & MAX Operations." *Proceedings of the Second International Fuzzy Systems Association*, Tokyo (July 1987): 827–30.

———. "Fuzzy Microprocessors—Rule Chip and Defuzzification Chip," *Proceedings of the International Workshop on Fuzzy Systems Applications*, Kyushu Institute of Technology (August 1988): 51–52.

Yun, H. *Lectures on Three Buddhist Sutras*. Taiwan: Fo Kuang, 1987.

Zadeh, L. A. "Fuzzy Sets." *Information and Control* 8 (1965): 338–53.

———. *Fuzzy Sets and Applications: Selected Papers*. Edited by Yager, Ovchnikov, Tong, and Nguyen. New York: Wiley, 1987.

INDEX